WHAT COUNTS IN TEACHING MATHEMATICS

Self-Study of Teaching and Teacher Education Practices

Volume 11

For further volumes:
http://www.springer.com/series/7072

WHAT COUNTS IN TEACHING MATHEMATICS

Adding Value to Self and Content

Edited by

Sandy Schuck
University of Technology Sydney, Sydney, Australia

Peter Pereira
DePaul University, Chicago, USA

Springer

Editors
Sandy Schuck
University of Technology Sydney
PO Box 222
Lindfield
NSW 2070
Australia
sandy.schuck@uts.edu.au

Peter Pereira
DePaul University
2320 Kenmore
Chicago IL 60614
USA
ppereira@condor.depaul.edu

ISSN 1875-3620
ISBN 978-94-007-3526-2 ISBN 978-94-007-0461-9 (eBook)
DOI 10.1007/978-94-007-0461-9
Springer Dordrecht Heidelberg London New York

Softcover reprint of the hardcover 1st edition 2011

Printed on acid-free paper

Springer is part of Springer Science+Business Media (www.springer.com)

Series Editor's Foreword

When the International Handbook of Self-Study of Teaching and Teacher Education Practices (Loughran, Hamilton, LaBoskey, & Russell, 2004) was launched at the American Educational Research Association Annual conference in 2004, the discussant, Denis Phillips of Stanford University posed an interesting question for the S-STEP community. Denis asked how the field might further develop and progress in the wake of the achievement of publishing a Handbook that so thoroughly captured the research literature and ideas around self-study. He was of the view that in publishing the Handbook the S-STEP community might feel as though the 'job had been done' and that furthering the core intent of self-study might be overtaken by an approach to gate-keeping that could limit the influence of self-study on the very field it was designed to foster – teaching and teacher education. This book by Schuck and Pereira, like that of Crowe's (2010) in Social Studies, offers another clear and strong response to Phillips' insightful challenge to the S-STEP community.

Schuck and Pereira have assembled an impressive group of mathematics educators, each of whom offers new ways of thinking about the teaching and learning of mathematics as a consequence of their involvement in self-study. The catalyst for their writing was a concern to "help our students develop mathematical pedagogical content knowledge as well as subject content knowledge … [because our] soon to be teachers of maths in primary and secondary schools need to recognise, and know how to reduce, the conceptual difficulties that often arise for school students". These authors' efforts in so doing are captured in the chapters of this book in ways that demonstrate an abiding commitment to teaching and learning about mathematics teaching that is the core business of mathematics teacher education.

The outcome of a serious consideration of a self-study methodology in researching the teaching and learning of mathematics is clearly evident in each of the chapters. Each of the authors demonstrates how they had to challenge their own conceptions of mathematics teaching and learning in order to develop deeper understandings of their own practice because, ultimately, they were concerned to genuinely challenge their own students' approach to, and understanding of, mathematics. It is this focus on their students' learning that has been so important in shaping their own learning about practice and is clearly an outcome derived from a thoughtful approach to researching practice through self-study.

Because self-study has been a guiding methodology to the research in this book, it is not surprising that other key aspects of teaching and learning also emerge. For example, the book is "loosely divided into two parts". The first part offers insights gained as a consequence of learning though mentoring and collaboration – a key aspect of self-study. The second part delves into the tensions and conflicts inherent in challenging students' attitudes and beliefs about mathematics and mathematics teaching. Through these two separate but closely related organizing principles, a coherent, thoughtful and honest account of salient issues in researching the teaching and learning of mathematics is portrayed for the reader in a most accessible and meaningful way. Schuck and Pereira have certainly assembled a core group of mathematics educators whose work demonstrates quality and scholarship in ways that offer many invitations and opportunities for others to build upon.

This book brings to the surface the value of learning with, and from, others through collaboration, mentoring and challenging the status-quo. Key concepts of reflection (Dewey, 1933) and framing and reframing (Schön, 1983) continually emerge in the chapters as the authors draw attention to the need to recognize and reconsider taken-for-granted aspects of practice that are so frequently overlooked or ignored in practice. In so doing, the ability to look again into the heart of teaching and learning in order to purposefully develop long-lasting pedagogical relationships in teacher education emerges as a theme that binds these chapters together into a coherent whole.

The editors describe self-study as "challenging, unsettling and uncomfortable". However, as they make clear, and as each of the chapters in this book more than illustrates, in accepting the challenge to look into practice in new ways, the sense of being unsettled and uncomfortable clearly leads to new learning that enhances the pedagogical experience for all of those involved. A crucial outcome of self-study is to build on and expand our knowledge of teaching and learning about teaching in ways that matter for the profession. In this book, Schuck and Pereira have done just that by drawing on the experiences of a range of authors who, through their common focus on mathematics, help to offer breakthroughs in a field in ways that genuinely model innovative approaches to studying mathematics teaching and learning.

This book is a fine addition to the growing body of literature of subject-specific self-study. It illustrates how understanding and responding to the self is important, but of itself, not sufficient for developing a pedagogy of teacher education (Korthagen, et al., 2001; Loughran, 2006; Ritter, 2007). Despite the superficial interpretations of those not involved in self-study but who often have opinions about the work as being too concerned with the self, this book makes clear that looking beyond the self is central to knowledge growth and development in ways that do make a difference for teaching and learning about mathematics teaching.

In a previous self-study Schuck (2009) highlighted the value of listening to, and learning from, her students of teaching. Likewise, Pereira (2005) has illustrated how to learn from the tensions and dilemmas associated with learning about how to become a teacher of mathematics. Together they have created a formidable editorial team and have developed a quality product that in years to come will be seen as a seminal work in the field of self-study of teacher education practices in mathematics.

References

Crowe, A. (Ed.) (2010). *Advancing social studies education through self-study methodology: The power, promise, and use of self-study in social studies education*. Dordrecht, The Netherlands: Springer.

Dewey, J. (1933). *How we think*. Lexington, MA: D.C. Heath and Company.

Korthagen, F. A. J., Kessels, J., Koster, B., Langerwarf, B., & Wubbels, T. (2001). *Linking theory and practice: The pedagogy of realistic teacher education*. Mahwah, NJ: Lawrence Erlbaum Associates Publishers.

Loughran, J. J. (2006). *Developing a pedagogy of teacher education: Understanding teaching and learning about teaching*. London: Routledge.

Loughran, J. J., Hamilton, M. L., LaBoskey, V. K., & Russell, T. (Eds.) (2004). *International handbook of self-study of teaching and teacher education practices*. Dordrecht, The Netherlands: Kluwer Academic Publishers.

Pereira, P. (2005). Becoming a teacher of mathematics. *Studying Teacher Education*, *1*(1), 69–83.

Ritter, J. K. (2007). Forging a pedagogy of teacher education: The challenges of moving from classroom teacher to teacher educator. *Studying Teacher Education: A journal of self-study of teacher education practices*, *3*(1), 5–22.

Schön, D. A. (1983). *The reflective practitioner: How professionals think in action*. New York: Basic Books.

Schuck, S. (2009). How did we do? Beginning teachers teaching mathematics in elementary schools. *Studying Teacher Education*, *5*(2), 113–123.

J. John Loughran

Contents

Contributors

Paul Betts Faculty of Education, University of Winnipeg, Winnipeg, Canada, p.betts@uwinnipeg.ca

Robyn Brandenburg School of Education, University of Ballarat, Ballarat, VIC, Australia, r.brandenburg@ballarat.edu.au

Nell B. Cobb DePaul University, Chicago, IL, USA, ncobb@depaul.edu

Joanne E. Goodell Cleveland State University, Cleveland, OH, USA, j.goodell@csuohio.edu

Sue Gordon The University of Sydney, Sydney, Australia, sue.gordon@sydney.edu.au

Hafdís Guðjónsdóttir School of Education, University of Iceland, Reykjavík, Iceland, hafdgud@hi.is

Jónína Vala Kristinsdóttir School of Education, University of Iceland, Reykjavík, Iceland, joninav@hi.is

Cynthia Nicol Department of Curriculum and Pedagogy, University of British Columbia, Vancouver, BC, Canada, cynthia.nicol@ubc.ca

Peter Pereira School of Education, DePaul University, Chicago, IL, USA, ppereira@condor.depaul.edu

Anne Prescott Centre for Research in Learning and Change, University of Technology Sydney, Sydney, Australia, anne.prescott@uts.edu.au

Sandy Schuck Centre for Research in Learning and Change, University of Technology Sydney, Sydney, Australia, sandy.schuck@uts.edu.au

About the Authors

Paul Betts is an Associate Professor at the University of Winnipeg. He taught mathematics in the public school system for 7 years before moving into a lecturer position at Brandon University in 1999. While at Brandon, he completed most of his PhD at the University of Regina in mathematics curriculum. He began a position at the University of Winnipeg in 2003 as an Assistant Professor specializing in mathematics education. In 2005, he successfully completed his PhD dissertation, which considered pre-service teachers' experiences with the nature of mathematics. He recently earned tenure and promotion at the University of Winnipeg. His current research focuses on (novice) mathematics teacher professional learning and the potential of philosophies of mathematics for the reform of mathematics education. p.betts@uwinnipeg.ca

Robyn Brandenburg is a Senior Lecturer in the School of Education at the University of Ballarat, Victoria, Australia. Her learning and teaching interests include a focus on developing pre-service teacher (PST) conceptual and mathematical pedagogical content knowledge, and reflective practice of teacher educators and PSTs. Her research is largely based on self-study methodology which is underpinned by systematically examining assumptions about learning and teaching as a means of enhancing pedagogy and practice. r.brandenburg@ballarat.edu.au

Nell B. Cobb is an Associate Professor of Mathematics Education at DePaul University, School of Education in Chicago Illinois. She is responsible for teaching middle and elementary mathematics education courses at the undergraduate and graduate levels. Nell also teaches content courses on mathematics for elementary teachers in the Mathematical Sciences Department. She is the co-director of the newly designed Masters of Science in Middle School Mathematics Education (MSME) Program and also serves as the Secondary Education Program Leader in the School of Education, Department of Teacher Education. Nell has coordinated the Algebra Project Teacher Resource Materials Team and she has coordinated a number of Algebra Project Professional Development Institutes. She has presented at several conferences and published articles in the areas of mathematics teacher competencies and general mathematics education. ncobb@depaul.edu

Joanne E. Goodell is an Associate Professor at Cleveland State University, where she teaches mathematics education and qualitative research methods courses, and directs the CSUTeach STEM teacher preparation program. She started her teaching career in Australia as a secondary mathematics teacher. Her research interests encompass self-study, gender issues and reform in mathematics education. j.goodell@csuohio.edu

Sue Gordon is Senior Lecturer at the Mathematics Learning Centre and Honorary Lecturer in the Faculty of Education and Social Work at The University of Sydney, Australia. Her teaching supports students' learning of mathematics and statistics at her university with a focus on statistics for Psychology. Her research is concerned broadly with teaching and learning in higher education including statistics education. Current projects include co-editing a Special Edition of the Statistics Education Research Journal on qualitative approaches in statistics education research, investigating university teachers' conceptions of student diversity in their classes and their views of effective teaching in a range of professional areas, and exploring the experiences of coordinators, teachers and students in mathematics bridging courses. sue.gordon@sydney.edu.au

Hafdís Guðjónsdóttir is Associate Professor in Special Education and Inclusive Pedagogy at the University of Iceland, School of Education. She graduated from Iceland Teacher College as a teacher in the year 1973. She earned a B.A. in Special Education from Iceland University of Education 1990, M.A. in Special Education from University of Oregon, Eugene 1993 and PhD in Special Education from the same school in the year 2000. For 25 years she taught as a general and special education teacher at compulsory schools. From the year 2000 she has taught at the University of Iceland School of Education. She has also collaborated with colleagues around the world, consulted with educators in Latvia, been a part of the European Agency for Development in Special Needs Education research group and the program chair for Self-study of Teacher Education Sig at AERA for 3 years. Her main research interest is inclusive education, pedagogy, teacher development and professionalism, teacher research, and self study of teaching and teacher education practices. hafdgud@hi.is

Jónína Vala Kristinsdóttir is an Assistant Professor of Mathematics Education at the University of Iceland, School of Education. She worked for 20 years as a general classroom teacher in an elementary school. Teaching is her primary profession and she has been active in curriculum planning and writing in mathematics for elementary schools. She teaches undergraduate and graduate courses on mathematics education and education in general. Her research interests are in mathematics learning and teaching in diverse classrooms. Current research and writing include teachers' development in mathematics teaching in diverse classrooms and a self-study of mathematics for all learners. joninav@hi.is

Cynthia Nicol is an Associate Professor of mathematics education in the Department of Curriculum and Pedagogy at the University of British Columbia, Canada. Her research focus is in the areas of teacher education, mathematics education, Aboriginal education, and culturally responsive research ethics and teaching practices. She is particularly interested in the methodologies of participatory action research, self-study research and community-based action research for the transformative possibilities they offer in researching mathematics, teacher and Aboriginal education. Her current research projects explore the nature of problem-based learning, culturally responsive education, and place-based learning. cynthia.nicol@ubc.ca

Peter Pereira (Editor), a retired Associate Professor in the School of Education at DePaul University in Chicago, has taught courses in mathematics, mathematics education, and curriculum. He is currently teaching a geometry course for middle school mathematics teachers focusing on developing mathematical habits of mind. He has published papers on deliberative curriculum theory, mathematics education, computer programming, dynamic geometry, and mathematics teacher development. He is on the international advisory board of *Studying Teacher Education* and his current scholarship concerns the complex process of becoming a mathematics teacher. He received a Master's degree from Harvard University and did his doctoral work at the University of Chicago. Prior to coming to DePaul, he taught mathematics in high schools in the United States and in England and was an instructor for mathematics teachers in India for two summers. ppereira@condor.depaul.edu

Anne Prescott is a Senior Lecturer at the University of Technology Sydney. Anne is the coordinator of the Teacher Professional Development Program in the Australian Centre for Child and Youth: Culture and Wellbeing and is a member of the Centre for Research on Learning and Change (L&C) in the Faculty of Arts and Social Sciences. Anne was a mathematics teacher in secondary schools before becoming a university lecturer. She coordinates the mathematics education subjects in the secondary education program at UTS. Anne's research interests are in student misconceptions, and the issues confronting pre-service and beginning teachers of mathematics as they enter the profession. anne.prescott@uts.edu.au

Sandy Schuck (Editor) is Associate Professor in Education at the University of Technology Sydney. Sandy is the leader of the research stream, Pedagogical Practice and Innovation, which is part of the University Research Strength, the Centre for Research on Learning and Change (L&C). She is the Coordinator of Education Higher Degrees by Research for the Faculty of Arts and Social Sciences. Sandy was a mathematics teacher in secondary schools before becoming a university lecturer. She coordinates the mathematics education subjects in the primary education program at UTS. Sandy's research interests are in three main areas: affective aspects of mathematics education, pedagogy with innovative technologies, and mentoring and induction of early career teachers. She has been engaged in self-study of her practices in teaching mathematics for over a decade. She is co-editor of two books on self-study of teacher education practices and has written numerous papers about her self-study work. sandy.schuck@uts.edu.au

Chapter 1
What Counts in Mathematics Education?

Sandy Schuck and Peter Pereira

Life is good for only two things: discovering mathematics and teaching mathematics.

Simeon Poisson

The Reform Movement in Mathematics Education

The authors of this chapter do not actually agree unconditionally with the above quotation, but we hope it has provoked a reaction from you, the reader. We want this book to be interactive; or as interactive as a book can be. By this we mean that we hope the book will challenge and provoke you, both in your thinking about the nature of mathematics and maths education, and in your reflections on your own teaching.

A recent article from an Australian newspaper asks the major parties in an election campaign, what actions will be taken to increase the number of students taking higher levels of maths. It goes on to decry the lack of qualified maths teachers in Australia (Joshi, 2010). Numbers of students taking up the study of higher maths at university have been steadily decreasing. On a personal note, typically when we confess that we are maths educators at social gatherings, the reaction is one of wariness, followed by confessions about a fear or loathing of maths at school. It is interesting to note that the two editors of this book live in two countries quite distant from each other. And yet, the problems and issues that are raised in discussions about maths and maths education fit as well in the educational debates of the one country as the other. These debates are centred on the nature of mathematical content, on the uptake of maths at higher levels, and on the reactions of the majority of the population to the learning of maths.

S. Schuck (✉)
Centre for Research in Learning and Change, University of Technology Sydney, Sydney, Australia
e-mail: sandy.schuck@uts.edu.au

S. Schuck, P. Pereira (eds.), *What Counts in Teaching Mathematics*, Self Study of Teaching and Teacher Education Practices 11, DOI 10.1007/978-94-007-0461-9_1,

While school mathematics is traditionally viewed as sets of rules, techniques, and algorithms unrelated to anything tangible and disconnected from everyday life, this is not how mathematicians view it. For them, the procedures are powerful tools that can help users to think about abstract ideas. Moreover, although mathematical concepts do have concrete referents and real-world connections, they gain their power precisely because they are abstract. Why, a mathematician might ask, waste time thinking about many specific applications when a single abstract formulation would suffice? This was the view taken by the "New Math" curriculum groups of the 1960s and 1970s. Dominated as they were by mathematicians, they aimed to produce students who would think as a mathematician would think (Meder, 1959, 1966). They believed that students would develop mathematically if they were exposed to mathematical structures and abstract mathematical ideas. Although this movement produced elegant materials and offered numerous summer courses and workshops, ultimately it was deemed a failure because of its concentration on abstract ideas and content (Kilpatrick, 1992).

A major new catalyst for change occurred in the late 1980s as concern grew about the lack of success experienced by students in their maths education. There was a change to an emphasis on problem-solving skills and on mathematical processes as well as content. The drive for change received added impetus when the influential American body, the National Council of Teachers of Mathematics (NCTM), published its first set of recommendations in 1989. These recommendations have subsequently been revised and have been accompanied by an impressive amount of support material (NCTM, 1989, 2000, 2006). The *NCTM Standards* include five "content" standards and five "process" standards describing the mathematical processes through which students should acquire and use their mathematical knowledge. The content standards spell out specific expectations (concerning number, algebra, geometry, measurement, and data analysis) for all grades, pre-K–12. The process standards direct mathematics teachers to place explicit emphasis on mathematics as a process of: *solving problems* rather than simply learning techniques; *reasoning* and constructing proofs rather than merely learning the appropriate reasons or examining proofs made by others; *communicating* with others to foster mutual understanding of mathematics; making *connections* to different mathematical ideas and to contexts outside of mathematics; and constructing many different *representations* of mathematical ideas to model a variety of phenomena. Although there are some dissenting voices, there is remarkable agreement amongst mathematics educators (and most mathematicians) in support of the emphasis on mathematical processes rather than an exclusive focus on abstract mathematical content. The *Standards* propose a new way of "doing" mathematics by focusing on authentic tasks, developing understanding, working collaboratively and problem solving.

Over the same time period, similar movements took place in many industrialised countries outside the USA. In Australia, for example, a national document was produced which supported the general approaches and underlying philosophy of the NCTM Standards (AEC, 1991). Its goals for students studying school mathematics were the following: to see the relevance, power and beauty of maths; to see maths as an activity involving representation and patterns; to use maths to empower them in everyday life and at civic and vocational levels; to be able to use the power of maths to present and interpret arguments, communicate and interpret results; to develop mathematical ways of thinking; and to understand the role maths plays in different cultures, both socially and technologically.

As with the *Standards*, the Australian document has been added to, modified and updated over recent years, with state governments in Australia producing syllabus documents which included sections on Working Mathematically (for example, NSW BOS, 2002) integrated with the mathematical content to be studied. The Australian National Curriculum, under development at present, indicates that the maths education curriculum for years K-10 should address the key concepts, skills and processes of maths, using approaches that ensure understanding, reasoning and problem solving, in a context of available digital technology (ACARA, 2009).

The Challenge for Maths Educators

Although mathematics education has been going through a prolonged period of intense reform as described above, it remains a subject that suffers from a lack of participation at the higher levels (see the discussion at the start of this chapter) and a lack of popularity at the lower levels. Primary school teachers and prospective teachers often are reluctant teachers and learners of maths, despite the best efforts of teacher educators to change this trend. Secondary maths teachers are often caught in a time warp in which approaches to teaching maths appear untouched by any reform movement. It is for this reason that self-study in maths education is so important. How can we innovate and energise the teaching of maths to student teachers so that they are, in turn, able to enthuse their students with a love for this subject and a confidence to engage in its intellectual pursuit? What kind of content should we be teaching and learning? What sort of pedagogical approaches are needed to change the way maths is perceived? In short, what are the roles and responsibilities of maths educators in teacher education programs?

We want to help our students develop mathematical pedagogical content knowledge as well as subject content knowledge. They need an understanding of the concepts that underlie basic maths skills, as well as a readiness to embrace higher maths. Our students, soon to be teachers of maths in primary and secondary schools, need to recognize, and know how to reduce, the conceptual difficulties that often arise for school students.

Educators, subject matter experts, politicians, and the general public often debate about the kind and amount of subject matter knowledge that teachers need. But since the mid-1980s, when Shulman introduced the term "pedagogical content knowledge" – abbreviated as PCK – to describe "ways of representing and formulating the subject that make it comprehensible to others" (Shulman, 1986, p. 9), the debate has intensified and the distinctions multiplied. Shulman himself used the same term in a subsequent article and made it clear that this is only one of at least seven different kinds of knowledge teachers should master (Shulman, 1987). PCK quickly became a widely used term with meanings that have been unpacked in various ways and applied to a variety of subject matters (See, for example: Ball, 1988, 1990; Bullough, 2001; Grossman, 1990; Gudmundsdottir, 1990; Wilson & Wineburg, 1988). In mathematics, Liping Ma – an American researcher fluent in Chinese – introduced the term "profound understanding of fundamental mathematics" (known as PUFM) to describe mathematics teachers' conceptual grasp of elementary mathematics. When comparing the knowledge held by mathematics teachers in the USA and China, PUFM best captured the differences she found and has subsequently become a widely used term (Ma, 1999). More recently, Deborah Ball, a member of the National Mathematics Advisory Panel, has asked specifically about what mathematical knowledge teachers need to know. Her research has further complicated the picture. In an empirical study of mathematical content knowledge for teachers (MKT) (Ball, Thames & Phelps, 2008), she distinguishes two different kinds of mathematical PCK (*knowledge of content and students* and *knowledge of content and teaching*) and two different kinds of "pure" content knowledge (*common content knowledge* and *specialized content knowledge*).

The task, then, for those of us who are maths educators is to teach prospective teachers not only how to teach content in schools but also how to teach processes. We strive to support students in working mathematically and in developing mathematical "habits of mind", that is to be critical thinkers, reflective about processes and content learned, and able to find and evaluate the most efficient strategies for solving problems. The challenge for teacher educators in maths education is consequently a large one. We are charged with "turning the tide", a responsibility not to be taken lightly.

There are two aspects of the approach taken in this book that, taken together, make it different from other discussions of mathematics education. First, we consider beliefs, attitudes and emotions, of both teachers and students, to be important drivers of successful mathematical experiences. Second, our approach to teaching maths is based on an understanding that both the affective and the cognitive sides of learning and teaching maths are inextricably intertwined. Thus, both play an important role in the learning and teaching of maths, and teachers need to be aware of both. Consequently, not only do we need to create positive and encouraging environments for students, we need to improve their attitudes and disrupt their beliefs about the nature of maths.

The Dual Nature of Educational Activity

> *The notion that the knowledge we seek to acquire exists in the outside world, independent of the wishes, beliefs, and intents of the knower, has been endemic in education. The attempt to draw a line between cognitive factual knowledge and subjective, so-called merely personal knowledge is not only artificial, but it also ends up distorting the realities of teaching and learning.* (Maxine Greene, 1977)

Most maths educators, including ourselves, are the products of a school system in which the cognitive side of maths was emphasised. Given that we chose to move into mathematical careers, we tend to be the students who benefited from the more traditional way of studying maths. In most cases, we were able to see its elegance and beauty without much prompting from our teachers. We enjoyed the order, logic and sequencing of the material. We did not experience the anxiety, fear and phobia that many others have when trying to learn maths. In many ways, this is a danger to our practice as maths educators. We do not readily appreciate the challenge that this deficit vision of maths presents for others. The reform movement in maths has been extremely powerful in alerting us to the deficiencies of more traditional maths education and has provided us with some useful alternatives.

We need to acknowledge that both the teaching and learning of mathematics, as well as of other subjects, has an emotional side. This has been recognized for a long time. Dewey (1930) frequently reminded us of the dual nature of educational activity, specifically describing the place of habits and impulses. Symonds (1943, 1944) examined relationships between the needs of teachers and their classroom behaviour. Jersild (1955) described the role emotions – especially anxiety, loneliness, hostility, and compassion – play in the life and work of teachers. In later work, Salzberger-Witenberger, Henry and Osborne (1983) examined the emotional experience of teaching and learning, Field, Cohler and Woole (1989) looked at emotions and learning from a psychoanalytic perspective, and Sarason (1993) talked of the importance of empathy in the preparation of educators. More recently, Sutton and Wheatley (2003) have published a review of the literature on the emotional aspects of teachers' lives and have indicated some directions for future research. Schuck and Grootenboer (2004) review the work done in Australia on the affective domain in mathematics education and indicate its importance in understanding mathematics education.

How Self-study Helps

It is here that self-study of teacher education practices becomes important. Most of us have had to challenge our own views about the accessibility of mathematical concepts and reframe our approaches to teaching maths. Self-study encourages us to constantly endeavour to improve our practice with a goal of supporting our students' learning and consequently their teaching of maths. The reflection in which we urge our students to engage is no less important for our practice. The need to be a critical thinker and learner applies to us as much as to our students.

Self-study encourages us to focus not only on our teaching but on our students' learning. It acknowledges the interlinking of our teaching and their learning. It also helps us to consider more than our personal experiences; to help to develop scholarship about teaching and teacher education. Self-study can be thought of as a contribution to scholarship; a way of understanding teacher education; and as a methodology for gaining insights into our practice that will contribute to the broader study of teacher education.

As maths educators, we are often caught between two central constructs. The one is the mathematical content to be taught and the other is our selves as teachers. The first is often thought to be objective in nature, the second subjective. In self-study we examine the objective and the subjective, the subject matter and the pedagogy, the content and the processes. We challenge our thinking about all of these with the aim of improving our practice (LaBoskey, 2004).

One of the critical contributions of self-study is the way it urges us to consider and reframe our practices and approaches. It is not easy to reframe our practices; this requires the uncovering of our assumptions and the challenging of our beliefs. We examine the living contradictions in our practice (Whitehead, 1989) and reframe our thinking (Loughran, 2002). Often we seek the support of others to help us challenge our assumptions. The others can be our students, the self-study literature, experts in the field and critical friends. Self-study is often best done with the help of others. Critical friends provide us with the opportunity to disrupt our beliefs in safe environments. They have the role of holding a mirror to us to show us what we cannot see on our own. As Shakespeare wrote:

> And since you know you cannot see yourself,
> so well as by reflection, I, your glass,
> will modestly discover to yourself,
> that of yourself which you yet know not of.
> Act 1, scene 2, Julius Caesar

Introduction to Chapters

In this book, the authors write about their experiences as maths educators. Most of the authors work in teacher education programs, teaching prospective secondary school or middle school teachers whose specialisation is in mathematics, or prospective primary school teachers who will be teaching maths along with a whole gamut of other subjects. One author works in a university learning centre supporting students in their studies of statistics. Her students are not prospective teachers but students from a variety of subject areas who are obliged to take a mandatory statistics course.

All of the authors have a passion for mathematics; they also have a deep concern for how mathematics is learned and taught. All of them strive to enhance the learning experiences of their students and to overcome the fear and resistance that many of their students display at the beginning of the course. They all engage in some form of self-study, some in more formal ways than others.

The book is loosely divided into two parts. The first includes self-studies that examine the nature of mentoring and collaboration with another. The critical friend is the focus of these studies. The second part examines self-studies which focus on interventions, processes of inquiry or investigation of practice that seeks to address problems in maths teaching and learning.

The authors come from a range of backgrounds as well. Some are novice or early career teacher educators, others are highly experienced educators who are eager to disrupt complacency. They come from a range of countries, but all seem to experience the same challenges in maths education. Four of the authors are Australian, one is Canadian, four are from the United States of America and two from Iceland.

The chapters about collaboration investigate what that collaboration brings to their self-study and to their understanding of maths teaching in particular. Betts investigates collaboration with novice teachers and examines the impact of his role as so-called expert in the collaboration; Guðjónsdóttir and Kristinsdóttir examine their collaboration as a maths educator and a special needs educator to see what each brings to the collaboration; Nicol looks at her work as a team member of a maths teacher education collaborative to identify ways in which this membership of the team enhances the learning of her students; and Schuck looks at her learning from a critical friend who is from a different discipline area, to ascertain what an outsider might see that an insider would not notice. For each of these authors the collaboration provides insights into their learning that they might not have experienced on their own.

The second set of chapters all investigate their authors' experiences of tensions and conflicts between their beliefs and attitudes about maths education and their students' beliefs and attitudes. What they have in common is a desire to support their students' learning of maths in a way that develops their habits of mind and improves their attitudes to the subject. Brandenburg examines how Roundtable Reflective Inquiries support student teachers as they develop understandings of their experiences as student teachers on practicum, and she learns about her assumptions as teacher educator. Cobb looks at her interactions with her students and considers what these tell her about her practice as a teacher educator. Goodell suggests ways that her practice supports the learning of her students. Gordon develops the idea of incongruities in her practice as she examines how she supports her students to learn statistics. Pereira and Prescott both look at the tensions created by the differing views about maths education that their students hold and that they hold. In their two chapters they discuss these tensions and how they try to resolve them through self-study.

Self-study is challenging. It can be unsettling and uncomfortable. At the same time, the benefits of self-study for enhancing our practice are substantial. We invite you to interrogate your own practices as you read this book and to reflect on the actions and thinking that the authors of the book have laid open for examination. We hope that you will find the book and your ensuing reflections worthwhile.

References

Australian Assessment, Curriculum and Reporting Authority (ACARA). (2009). Phase 1: The Australian Curriculum. Accessed August 19, 2010, from http://www.acara.edu.au/phase_1_-_the_australian_curriculum.html

Australian Education Council. (1991). *A national statement on mathematics for Australian schools*. Melbourne: Curriculum Corporation.

Ball, D. L. (1988). *The subject matter preparation of prospective teachers: Challenging the myths*. East Lansing, MI: National Center for Research in Teacher Education.

Ball, D. L. (1990). The mathematical understandings that prospective teachers bring to teacher education. *Elementary School Journal, 90*, 449–466.

Ball, D. L., Thames, M. H., & Phelps, G. (2008). Content knowledge for teaching: What makes it special? *Journal of Teacher Education, 59*(5), 389–407.

Bullough, R. V. (2001). Pedagogical content knowledge circa 1907 and 1987: A study in the history of an idea. *Teaching and Teacher Education, 17*(6), 655–666.

Dewey, J. (1930). *Human nature and conduct*. New York: Random House.

Field, K., Cohler, B., & Woole, G. (1989). *Emotions and learning: Psychoanalytic perspectives*. Madison, CT: International University Press.

Greene, M. (1977). Toward wide-awakeness: An argument for the arts and humanities in education. *Teachers College Record, 79*(1), 110–125.

Grossman, P. L. (1990). *The making of a teacher: Teacher knowledge and teacher education*. New York: Teachers College Press.

Gudmundsdottir, S. (1990). Values in pedagogical content knowledge. *Journal of Teacher Education, 41*(3), 44–52.

Jersild, A. (1955). *When teachers face themselves*. New York: Teachers College Press.

Joshi, N. (2010). Wrong numbers. The Australian newspaper. Accessed August 9, 2010, from http://www.theaustralian.com.au/national-affairs/commentary/wrong-numbers/story-e6frgd0x-1225902743924

Kilpatrick, J. (1992). A history of research in mathematics education. In D. Grouws (Ed.), *Handbook of research on mathematics teaching and learning*. Reston, VA: NCTM.

LaBoskey, V. K. (2004). The methodology of self-study and its theoretical underpinnings. In J. Loughran, M. L. Hamilton, V. K. LaBoskey, & T. Russell (Eds.). *International handbook of self-study of teaching and teacher education practices* (pp. 814–817). Dordrecht: Kluwer.

Loughran, J. (2002). Effective reflective practice: In search of meaning in learning about teaching. *Journal of Teacher Education, 53*(1), 33–43.

Ma, L. (1999). *Knowing and teaching elementary school mathematics: Teachers' understanding of fundamental mathematics in China and the United States*. New York: Lawrence Erlbaum Associates.

Meder, A. E. (1959). Sets, sinners, and salvation. *Mathematics Teacher, 52*, 434–438. Reprint: (1966). *59*, 358–363.

NCTM. (1989). *Curriculum and evaluation standards for school mathematics*. Reston, VA: NCTM.

NCTM. (2000). *Principles and standards for school mathematics*. Reston, VA: NCTM. Available at http://www.nctm.org

NCTM. (2006). *Curriculum focal points*. Reston, VA: NCTM. Available at http://www.nctm.org

NSW Board of Studies (NSW BOS) (2002). *Mathematics K-6 Syllabus 2002*. Sydney: Board of Studies NSW.

Salzberger-Witenberger, I., Henry, G., & Osborne, E. (1983). *The emotional experience of teaching and learning*. London: Routledge.

Sarason, H. B. (1993). *The case for change; Rethinking the preparation of educators*. San Francisco: Jossey-Bass Publishers.

Schuck, S., & Grootenboer, P. (2004). Affective issues in mathematics education. In B. Perry, G. Anthony, & C. Diezmann (Eds.), *Research in mathematics education in Australasia 2000–2003* (pp. 53–74). Flaxton, QLD: PostPressed.

Shulman, L. S. (1986). Those who understand: Knowledge growth in teaching. *Educational Researcher, 15*(2), 4–14.

Shulman, L. S. (1987). Knowledge and teaching: Foundations of the new reform. *Harvard Educational Review, 57*, 1–22.

Sutton, R. E., & Wheatley, K. F. (2003). Teachers' emotions and teaching: A review of the literature and directions for future research. *Educational Psychology Review*, *15*(4), 327–358.
Symonds, P. (1943). The needs of teachers as shown in autobiographies. *Journal of Educational Research*, *36*, 662–677.
Symonds, P. (1944). The needs of teachers as shown in autobiographies; II. *Journal of Educational Research*, *37*, 641–653.
Whitehead, J. (1989). Creating a living educational theory from questions of the kind, 'how do I improve my practice?' *Cambridge Journal of Education*, *19*(1), 41–52.
Wilson, S. M., & Wineburg, S. S. (1988). Peering at history through different lenses: The role of disciplinary perspectives in teaching history. *Teachers College Record*, *89*(4), 525–539.

Part I
Collaborations and Critical Friends

Chapter 2
Tensions of Mentoring Mathematics Teachers: Translating Theory into Practice

Paul Betts

Introduction

I am a junior academic working with teachers as part of several participatory action research projects. These projects aim to develop a mathematics teacher professional learning research agenda. Teachers, most of whom are novices, invited me into their classrooms as we sought to develop, implement and reflect on reform-based mathematics teaching strategies. In turn, I invited them into my research activities as we sought to notice our collective and individual professional learning. Our roles blended.

And yet, I was an outsider – the mathematics educator, not the teacher – "having" an expert knowledge of mathematics for teaching. I could not escape this positioning, as much as I wanted to be viewed as a learner with the teachers, and as much as I valued each teacher's inquiry into their own classroom activities. How should I respond to teachers, especially novices, as they positioned me as a mathematics education expert who could answer their questions? I wondered how to fill this role.

So I took on the role of mentor. A mentor–mentee relationship is dynamic, not one-way, not an assigned role; rather, it is a space for dialogue that seems to be occasioned by a perceived differential knowledge base. The teachers did position me as "having" extensive knowledge of mathematics for teaching, and I often obliged with explicit didactical comments in response to their questions. But the relationship also carried an opportunity to notice qualities of "mathematical knowledge for teaching" (Ball, Thames, & Phelps, 2008), that is, knowledge of mathematics used by teachers when teaching mathematics, both mine and those of the teachers with whom I worked.

I seek to illuminate the tensions of a rookie researcher, as I shifted my theoretical understandings into the practice of working closely with teachers in collaborative

P. Betts (✉)
Faculty of Education, University of Winnipeg, Winnipeg, Canada
e-mail: p.betts@uwinnipeg.ca

S. Schuck, P. Pereira (eds.), *What Counts in Teaching Mathematics*, Self Study of Teaching and Teacher Education Practices 11, DOI 10.1007/978-94-007-0461-9_2,

research/teaching contexts. All the literature I have read on research methods, mathematics education and professional development did not prepare me for the roles I would play in this work. I claim that mentoring within a collaborative learning environment is a highly dynamic positioning and that it is impossible to fully pin down effective mentoring. I also claim that seeking a mathematical knowledge for teaching is both helpful and problematic. There will always be tensions when being a mentor and when using mathematical knowledge for teaching, and it is a mistake to ignore these tensions.

In this chapter, I first lay out some theoretical issues of relevance concerning both mentoring in a collaborative environment and mathematical knowledge for teaching. This "theory" is positioned as problematic to set the stage for noticing tensions within my own practice. Three episodes are described to illustrate the tensions that did emerge from my experiences. In the discussion, I draw themes from the episodes, which suggest that these tensions can arise for others doing work with teachers. Finally, I conclude with some reflections concerning the challenges of mentoring and mathematics teacher education.

Theoretical Framework

As a rookie researcher, I entered the work of collaborating with teachers with plenty of book knowledge; I was theoretically prepared. I had considered literature concerning the professional development and learning of teachers; both how teachers learn and develop, and what could/should be learned by mathematics teachers. Difficulties arose for me when I used this theoretical knowledge in practice. It is not a matter of simply applying theoretical principles to the practice of working with teachers. In what follows, I describe some of the work I have read, and note ways that these theories could be problematic. This will set the stage for the tensions I experienced in my work with teachers.

Mentoring Within a Collaborative Learning Environment

I view collaborative teacher research as characterized by the notion of inquiry, where teachers investigate their own teaching agendas within their own practice (Levin & Rock, 2003). Collaborative learning environments are effective when they are both safe and critical (Darling, 2001). That is, the community seeks to include and validate the voices of all participants, to be nurturing and supportive, while seeking to critically and constructively engage with the problems of teaching. As a member of these communities, I tried to begin with the needs of the teachers; to validate their needs, ideas, and expertise; to work with them in their classrooms; to structure opportunities for a group of teachers to reflect on their own classrooms; to co-enquire with teachers into the teaching and learning of mathematics. In short, I tried to be the catalyst by which a safe and critical community emerged.

It is generally agreed upon (Bullough & Draper, 2004) that mentoring involves a relationship between two people with differing amounts of experience in a profession, the mentor being more experienced but still benefiting from the relationship, although in different ways from the mentee (Jeruchim & Shapiro, 1992). This relationship may take the form of an experienced teacher working with a novice or pre-service teacher. The notion of expertise follows from the greater experience of the senior teacher. For me, as a professor working at a university, my expertise is somewhat different from the experienced teacher as I do not (or no longer) draw extensively from practice but rather from a theoretical base. Working with teachers is an opportunity for me to benefit from these relationships by reconstructing practical understandings of a theoretical knowledge base.

It is also well established that the quality of any mentor–mentee relationship emerges from a trusting relationship (Jeruchim & Shapiro, 1992; Halai, 1998; Morton, 2005; Portner, 2008). Mentors can work toward a quality relationship by adopting various dispositions, such as teacher (Halai, 1998; Maynard & Furlong, 1993), coach (Halai, 1998; Portner, 2008), guide (Portner, 2008), and co-enquirer (Maynard & Furlong, 1993). In order to foster a facilitative and trusting relationship, Grisham, Ferguson and Brink (2004) recommend that mentors adopt a position of low direction and high support. All of these dispositions are affirmed, explicitly or implicitly, in The National Council of Teachers of Mathematics (NCTM) recent publication for mentors of beginning mathematics teachers (Zimmermann, Guinee, Fulmore, & Murray, 2009).

To highlight how a mentoring role could be enacted within a collaborative environment, I focus on a particular model of potential mentor dispositions. Based on mentor relationships between experienced and pre-service teachers, Young, Bullough, Draper, Smith and Erickson (2005) found three general mentoring patterns, namely responsive, interactive and directive. The interactive pattern values a dialogic relationship, whereas the responsive mentor seeks to always respond to the needs of the mentee, and the directive mentor sets the agenda for what will be learned. This model places mentor disposition on a continuum from highly hands-off to highly authoritarian.

Because my work with teachers is collaborative, I sought to find balance between responsive and interactive dispositions. I sought to be responsive by encouraging teachers to decide on the direction of our research and by providing various supports and resources. I sought to be interactive by trying to position myself as a learner along with the teachers, and by validating the differing understandings that each of us could bring (Portner, 2008). Similar to the experiences of Halai (1998), the teachers (especially novices) positioned me as a "problem solver or answer giver, as if I had all the answers or all their problems had solutions" (p. 303). At times, I found myself sliding into a directive stance as the teachers sought answers from me. This directive pattern was more than just providing resources, support and guidance, as might be expected given a responsive pattern. Rather, I felt myself wanting to tell teachers (especially novices) what to do, while also wanting to resist being positioned as a bastion of knowledge.

The work of mentoring within a collaborative learning environment with teachers could be problematic in three ways. First, I privilege trust building over critique, wanting to "nurture personal and professional growth of the participating teacher[s]" (Halai, 1998, p. 298). But collaborative learning environments should also be critical. Second, consistent with reform agendas, I see my role as facilitating the professional learning of teachers. But often, the novice teachers that I worked with positioned me as answer giver, and I found myself telling; I found myself shifting from a responsive or interactive to a directive disposition. Finally, being positioned as expert, and bringing largely theoretical understandings to the community, is at odds with my desire to learn, to develop understandings of mathematics education grounded in praxis.

Mathematical Knowledge for Teaching

The question of what teachers need to know is drawing considerable interest from researchers. Since the work of Shulman (e.g., 1987) in the late 1980s, who shifted attention toward the notion of pedagogical content knowledge (PCK) – a recognition of the knowledge of teachers that blend the subject to be taught with pedagogy – various researchers have tried to unpack the PCK of teachers in various subjects. Within mathematics education, Ball and her colleagues have done considerable work to categorize the multifaceted mathematical knowledge of teachers. A recent special issue of the journal *For the Learning of Mathematics* (Volume 29, number 3, November 2009) speaks of the current importance in mathematics teacher education of understanding teachers' knowledge of mathematics as it pertains to their classroom practice.

Ball and her colleagues extended the work of Shulman by closely examining the practice of teachers. Mathematical Knowledge for Teaching (MKT) is "the mathematical knowledge needed to carry out the work of teaching mathematics" (Ball et al., 2008, p. 395). This is the knowledge often used by teachers when performing any element of mathematics instruction, including planning and implementing activities, assigning homework, assessment and evaluation of student work, meeting with parents, and working within the local context of the school.

Ball et al. (2008) illuminate MKT by considering the sorts of thinking and behaviours of teachers that have mathematical import. To that end, they distinguish between common content knowledge (CCK) and specialized content knowledge (SCK). CCK is knowledge of mathematics that arises in many contexts, such as the standard algorithm for adding two numbers. SCK is mathematical knowledge that only teachers need to know, such as determining if a student's nonstandard approach to adding two numbers is mathematically valid. The emergence of SCK involves an "unpacking" or "decompression" of mathematics by teachers, in ways needed for effective teaching but not needed by the average person. Such distinctions reconstruct teacher knowledge of mathematics as a domain of mathematical knowledge. Further, rich articulations of MKT can provide guidance when considering mathematics teacher education questions.

The notion of MKT and Ball's refinements could be problematic in terms of my collaborative work with teachers. First, I want to value the work of teachers, as a way to build relationships with teachers. But MKT brings forth the noticing of deficiency within teachers' understandings of teaching mathematics. For example, a recent article by Hill and Ball (2009) opens with the story of Ms. González, a teacher using black and red chips to model addition of integers. When the teacher encounters a subtraction question [e.g., $(-1)-(-3)$], she is unable to represent the problem with the chips, and resorts to an algorithmic explanation [i.e., when subtracting integers, add the opposite: $(-1) - (-3) = (-1) + 3 = 2$]. Hill and Ball use this example to illustrate a deficiency in the specialized content knowledge of Ms. González. I wonder how Ms. Gonzáles would feel about the use of MKT to analyze what she does not understand adequately. At times, I too have felt myself lacking a rich enough MKT, despite extensive mathematical background, theoretical study and experiences working with teachers and children. This leads me to question whether MKT can be used as more than an analytical research tool. Is it a tool for critique by reflective teachers? Or perhaps it operates as a dagger that kills further analysis of teaching by teachers because it points at deficiency and triggers emotional barriers to learning.

Second, a concern for the qualities of collaborative learning environments leads to concern with a universal rendering of MKT. On the one hand, it seems obvious that a richer and multidimensional knowledge of mathematics for teaching leads to more effective mathematical learning environments. On the other hand, it is not so clear what MUST be part of the knowledge base, or how this knowledge is developed. Mathematics educators probably could agree on some core principles, but some "pieces" of knowledge might be considered of less importance, secondary, or even esoteric, or could emerge from teacher knowledge outside of MKT. For example, Hill and Ball (2009) suggest that Ms. González (see previous paragraph) might have responded with a conceptual explanation if she knew the deeper connection between a "taking away" representation of integer subtraction and "regrouping" when subtracting multidigit whole numbers. But knowing such a deeper connection is not necessary knowledge to respond appropriately. Ball and her colleagues also provide compelling evidence that teachers must be able to evaluate the mathematical validity of student solutions, often on the fly, which suggests teacher education curriculum include opportunities for teachers to evaluate the actual work of students. But a teacher with an open disposition toward teaching could find ways to respond appropriately to student work, without actually fully unpacking its mathematical validity. Such an experience is likely to contribute to richer MKT, especially given reflection on such an episode by the teacher, while not being something that MUST be learned during a formalized teacher education curriculum. Examples of student work, although generated by actual teaching practice, may turn out to be rare within the actual teaching experience of a particular teacher. Or a teacher may carry esoteric pieces of MKT that are generally not needed by other teachers. The notion of one MKT for all becomes problematic as I try to work with teachers and participate in a supportive, trusting, critical and reflective learning environment.

Three Episodes from My Practice

I believe the problematic aspects of theory described in the previous section underlie the tensions that emerged in my work with teachers. The following episodes illustrate many tensions: When do I tell, interact, respond? When do I critique, support, coach, guide, instruct? When am I learner, mentor? Will MKT be a barrier or a tool? When do I admit to inadequacy? In each of the three episodes below, I explore how I faced these questions and navigated my tensions.

Mentor Telling Mentee and Growth of a Novice Teacher's MKT

I am working with a novice teacher (Susan) and an experienced teacher (Cam) at a school whose grade 5–8 teachers are working to take up reform-based approaches to teaching mathematics. Susan is comfortable with reform approaches, based on her experiences as a pre-service teacher – her practicum placements were in schools that valued inquiry approaches, and her math methods courses emphasized similar approaches (I was her instructor). She finds herself in a school where her colleagues and her grade 8 students are not familiar with these approaches to teaching. She is struggling to adjust to the challenges of her students to the point where finding ways to successfully (read, "smoothly") implement methods learned during her pre-service education is beyond her ability. Susan is also happy to work with Cam, who is viewed as an effective teacher and is also teaching grade 7 and 8. Cam draws on reform-based approaches such as using manipulatives and engaging students with math by "making it real". Two grade 5 and 6 teachers are part of our collaborative learning group, and we have agreed to work on developing the problem-solving skills of their students. Usually, I work with the teachers in pairs based on grade level to plan, implement and reflect on lessons. The story below considers a planning session to implement the "King Arthur Problem" (found in Burns, 1996) with Susan's class, where Susan would play the lead and Cam and I would help.

I consider the King Arthur Problem to be a high-level task. Such tasks are layered with multiple possible responses, thus providing a context for inquiry, thinking, reasoning and problem solving (Smith, Hughes, Engle, & Stein, 2009). Cam is worried about whether the students will be engaged by the problem, clearly not worried about content. Susan, on the other hand, appears focused solely on content as she worked toward understanding several possible ways to solve the problem. I facilitated Susan's exploration by sharing other possible solutions. But then she asks: "I need to know what to do when a student gets stuck". A little discussion reveals that Susan is trying to anticipate/plan for the scaffolds she will need to help her students be successful. Her inquiry started in content (ways to solve the problem), but clearly moved into a pedagogically grounded understanding of the layers of mathematics possible within the problem – that is, the need to help students move through the problem in their own ways. Cam quietly watched the entire dialogue. Susan kept asking and I kept answering with explicit possible scaffolding questions.

Throughout this inquiry with Susan and Cam, I felt conflict between directive and responsive dispositions. I faced the question of how to respond as a mentor – the tension was between telling her possible scaffold questions and facilitating her ability to further unpack the problem and make her own moves from content to pedagogy. Susan was responding from her inexperience as a teacher, wanting (needing?) a tighter plan to help her negotiate the open-ended problem-solving lesson we were developing, which was largely foreign to her past experiences. This is the teacher's inquiry, so would telling her answers disable her ability to transfer the skill to other problems and contexts? Susan appeared un-phased by my answers. She insisted and would expect nothing less than explicit possible scaffold questions for each stage of the problem. The irony of Susan asking for scaffolds and me not scaffolding was not lost on me.

The tension was furthered by the presence of an experienced teacher with, apparently, a completely different agenda or need – how to present the problem in a way to engage students. I felt watched by Cam, even as I thought "facilitate, don't tell" and continued to tell Susan possible scaffold questions while she wrote them down. In that moment, did I respond with explicit and direct answers because the expert teacher was watching, not just because the novice was asking and positioning me as a source of answers? Was I wondering if (hoping that) Cam saw me as having expertise in teaching math, not just a university professor who lives in an ivory tower?

I am not suggesting that Cam responded with silence because he was an experienced teacher and Susan responded with questioning because she is a novice teacher. Rather, I faced the problem of trying to help a novice teacher to survive. Teachers continually make decisions on the fly, sometimes by habit and sometimes explicitly. Susan was overwhelmed by the number of decisions she needed to make on the fly. Our dialogue was grounded in the need to curtail the amount of on the fly decision-making needed during the lesson. My tension of facilitate versus tell was navigated by doing both of these. Our dialogue appeared directive on the surface but it was still critical and dialogic. We sought to engage with the problem of doing inquiry math with students who find this method of teaching foreign, with an experienced teacher who may not have fully perceived the nature of inquiry available with the King Arthur problem, and with a novice teacher facing the challenges of first year. The dialogue was critical because we faced legitimate and real problems of teaching.

What emerged from our critical dialogue was growth in Susan's MKT by a blending of prior knowledge grounded in her personality and need. Susan's CCK is already strong, and her general pedagogic knowledge is theoretically strong but still emerging in practice. The growth in her MKT is driven by a need to survive the first year of teaching and maintain her sense of what it means to teach (read, inquiry is the way to go). Susan started to recognize the need to anticipate possible student difficulties and plan possible responses to these students. She shifts from how to solve the problem to how to navigate various and possible student problem-solving trajectories. She is seeing the relationship between the rich mathematics involved and teaching behaviours that foster inquiry. Susan's personality and need drive a blending of CCK and general pedagogic knowledge, resulting in enhanced understandings of teaching. Susan's MKT grows because the dialogue is critical, and not because of my telling behaviours as a mentor.

Mentoring Enhances My MKT

At another school, I worked with Amy, a beginning teacher of grades 2 and 3, who found herself worrying about the more traditional teacher in the next grade, and yet was also embracing the idea of providing opportunities for students to develop their own understandings. Amy and I decided to design activities where students explore the notion of area as covering. In one activity, we asked students to generate as many shapes as they could with an area of 16 square units, by drawing shapes on graph paper and counting the number of grid squares inside the shape. We expected students to consider shapes that were composites of rectangles, such as an "L" shape (two rectangles butted together at 90°). Thus, we were pleased when many of the students started making block letters with the desired area, and we challenged them to make every letter in their name. Part way through the activity, one student asked if he could try a circle. This request was unexpected, and Amy faced a decision, "If I don't allow circles, this student might become disruptive. But circles will be hard because of coordinating the partial grid squares from a curved edge". Amy allowed the student to consider circles. It wasn't long before several students were looking for cups and other objects that could be traced to form a circle on graph paper. Near the end of the activity another student, Trevor, noted, "it is hard to get close to sixteen because when I try a different sized can, the area changes a lot". Amy was focused on teaching issues such as how to wrap up the activity because time was short. I followed her lead, and we didn't notice in-the-moment something mathematically wonderful.

Later, while reflecting together, I asked Amy why she allowed the student to consider circles, and she revealed her concern with the student becoming disruptive. Not wanting to get bogged down in classroom management issues, I suggested we focus on the mathematics for now. Amy agreed, and I started talking about the area of a circle formula and said something I knew formulaically: "The area is related to the radius squared. This means that if you double the radius, the area quadruples". Amy nodded agreement. Having majored in mathematics at university, Amy is comfortable discussing symbolic formulas. I was just thinking out loud and didn't know where my line of thought would take me. I continued talking: "That is what was happening with the containers. That little change in radius caused a larger than expected change in area, perhaps because the children carried linear intuitions about how area would change. That explains Trevor's comment" [see above]. At that moment I was having an epiphany about the meaning of the formula for the area of a circle based on observing the thinking of a student. Amy, on the other hand, was confused by my comment, focusing rather on her decision to allow circles because of classroom management issues. So I went through it again, and my formulaic understanding shifted and consolidated into a deeper conceptual understanding of the formula for area of a circle.

Amy nodded her head. This insight about the formula for the area of a circle was new to her. She framed my explanation as grounded in expertise in children's learning of mathematics. She didn't hear in my voice the shift in my own understanding, the import of my excitement. And I did not have the courage to tell her it was new for me too, in a way: a shift from "formula implies double r = quadruple A" to "small changes in radius trigger larger than linear (and larger than expected from the perspective of a learner) changes in area".

It is clear from the story above that extensive education (read, training?) in mathematics at the university level is insufficient knowledge for the tasks of teaching, a claim that would be supported by Ball and her colleagues (2008) and others (e.g., Borko et al., 1992; Meredith, 1995). Borko et al. (1992), for example, described a student teacher with extensive background in mathematics who could not explain conceptually the algorithm for dividing two fractions. Amy and I both majored in mathematics for our undergraduate degrees, and yet we both carried formulistic understandings of the area of a circle. I completed a masters in mathematics and then taught mathematics for several years, but it was not until my PhD studies that I started learning representations of arithmetic procedures, knowledge which deepened my conceptual understandings of standard algorithms. Only recently did I develop a repertoire of understandings of dividing fractions so that I feel comfortable explaining to teachers why the division algorithm for fractions works. Ball and her colleagues would count these examples as evidence that teachers need more than just common content knowledge of mathematics (e.g., how to perform standard algorithms); teachers also need specialized content knowledge such as being able to explain why you invert and multiply to divide fractions and, as in this episode, why the area of a circle is not linearly related to the radius.

Although this story illustrates aspects of MKT, it also illuminates tensions concerning the necessary MKT of mentors. The above story is an example of Amy and my further "unpacking" the formula for the area of a circle. But this unpacking is the happy circumstance of three accidents: first, a student decided he wanted to consider circles, which we had not planned for; second, another student observed and vocalized larger than expected changes in area of circles; and third, we had to notice and make sense of this student's informal observations, which hinged on me not wanting to get side tracked by classroom management issues. Without these three accidents, we would not have unpacked the formula for area of a circle. I can only wonder in what other ways my MKT can be deepened to be needed/useful in my work with teachers. I am delighted that my MKT was deepening, but disturbed it occurred accidently.

I also felt tension as a mentor. Amy has continually positioned me as answer giver and expert, as if I know more and can answer all her questions. Others have reported that mentors learn from their mentoring experiences, but in different ways from their mentees. Here I was learning the same thing as Amy! I was learning something that Amy assumed I already knew; something that would likely be considered a valued datum of MKT; something I should already know to work with teachers.

Should I have admitted that it was new for me too? Would this have mattered for the professional learning of Amy – would it perhaps be a validation of new insight despite being mathematics majors; or perhaps a wondering about the expertise of the mentor? Did it matter that I tried to position myself as a learner with the teachers I work with? Perhaps my disposition was a necessary condition for noticing and making sense of a student's observation about circle areas.

Tensions Within My MKT: The Story of Subitizing

In order to set the stage for my next story, I need to delineate between two discourses concerning subitizing, that of teachers and math consultants in my province, and of psychologists. Subitizing is "instantly seeing how many objects" there are in a set. It is distinct from the slower process of counting (Clements, 1999). Psychologists distinguish between these two processes because of observations of the ability of all humans to very quickly enumerate small (*randomized*) collections of objects (Dehaene & Cohen, 1994). Cognitive models of mathematical development include subitizing as a critical competency because of evidence suggesting that the ability of young children to subitize correlates with later maths ability (LeFevre et al., 2010). In my province, the prevailing view seems to be that children should be provided opportunities to improve their ability to subitize. Activities in widespread use involve quickly recognizing the quantity of a *familiar pattern* (e.g., the five on a die is always four corner dots and one in the middle), using flash cards, ten frame cards, dominoes or dice.

Students in my elementary math-teaching methods course frequently ask me about subitizing, having observed teachers discuss the idea. My response to students is based on the cognitive psychology literature: there is no evidence to support the claim that subitizing can be taught (it is likely an innate skill of all humans), and the activities described in the previous paragraph are not subitizing. On the other hand, recognizing the quantity of familiar arrangements of objects is helpful for occasioning the development of young children's number sense. I felt it was important for pre-service teachers to be aware of two discourses surrounding subitizing, one grounded in the practice of teachers and another in the work of psychologists. I would say that students perceive me as having highly critical and aggressive views on subitizing.

And now the stage is set for a very short story. Helen is a beginning teacher, with whom I work, who took my pre-service math teaching courses and who has heard my views on subitizing. Cynthia is an experienced teacher, with whom I also work, who has not heard my views on subitizing. Cynthia, Helen and I worked together to learn about developing the number sense of grade 1 children. On one occasion, the issue of subitizing arose, and Cynthia accurately reproduced the prevailing view in the province. I said nothing, and Helen looked at me questioningly. I believe Helen remembered that my views were critical of the prevailing view and was wondering why I didn't say anything. The conversation moved on to other issues, and we did not return to subitizing.

In my later reflections, I wondered why I was silent at that moment. I deliberately work to always be respectful of the work and understandings of teachers, so perhaps my silence was an act of respect. I recognize the importance of relationships in my work, so perhaps I was worried that stating my views would introduce a negative element into our group. Could I find myself ineffectual because Cynthia positioned me as adversarial and challenging rather than respectful of teachers? Would it matter to our collective and individual professional learning to introduce an alternative view of subitizing? And I worried what Helen thought of me and the situation – how would she view my silence and would it matter to her view of subitizing?

This story led me to seek further literature concerning mentoring and concerning subitizing. I first consider mentoring. Orland-Barak (2005) asserted that:

> mentoring seems to be strongly shaped by a struggle between competing discourses, whereby mentors often find themselves lost in trying to translate one discourse of practice into another. The result is that, often, mentors find themselves "speaking one language" and practicing another one. (p. 364)

Orland-Barak called this disjuncture "lost in translation", and the concept seems to apply to my situation. I adopted a critical stance toward subitizing with pre-service teachers and a silent one with an in-service teacher. I failed to translate my understandings to a mentoring context. It is not enough to embrace a responsive, dialogic or directive disposition as a mentor. Nor is it enough to take up differing dispositions as the context may seem to suggest, such as directive when the mentee seems to need rescuing and responsive when the mentee seems more able to bring forward their own learning agenda (Young et al., 2005). The mentor must also translate their understandings from one practice to another – in this case, how might my understandings of subitizing translate when working with experienced teachers with differing perceptions of the importance of subitizing.

When I looked again at the subitizing literature, seeking to better understand the concept of subitizing, I found two competing discourses. On the one hand, there was the perspective of psychologists described above. On the other hand, mathematics educators take a different view, distinguishing between two kinds of subitizing, perceptual and conceptual. Drawing on Clements (1999), perceptual subitizing is "recognizing a number without using other mathematical processes" (p. 401), which roughly maps onto the definition used by psychologists, whereas conceptual subitizing is instantly knowing a quantity by recognizing a "number pattern as a composite of parts and as a whole" (p. 401), such as the five on a die is four dots in the corner and one in the middle. Conceptual subitizing maps onto the subitizing endorsed by math consultants and elementary teachers in my province. I had a blind spot, generated by attending only to randomized pictures of quantity, as in the psychology literature.

As it turned out, my silence, based on respect and open mindedness, was a reasonable response to experienced teachers who endorse the importance of teaching subitizing. I do not need to spend time with them noting the difference between conceptual and perceptual subitizing, but I probably do need to soften and elaborate on my approach with pre-service teachers when they ask about subitizing during my math methods courses.

This may seem like a clean conclusion to this story, but there are still two tensions for me that remain unresolved. First, although subitizing seems to be a possible MKT datum, how in depth an understanding is needed by teachers? Is it enough that teachers attend to a conceptual understanding of quantity by recognizing the important role of pattern recognition, and thus providing opportunities for children to determine quantity by patterning? Or, at the other extreme, should teachers be made fully aware of the lines of research in psychology and education, thereby seeking to draw implications for learning and their own teaching practice? The answer to this question is likely context dependent and therefore not possible to resolve in advance, hinging on how each teacher responds to and takes up the idea that there is more to determining quantity than counting.

Second is my tension as a mentor. Pre-service teachers also see me as an expert and answer giver, and I try to respond sincerely. But, I found my strong views on subitizing needed refinement. Have I led my pre-service teachers astray? What other areas of my knowledge of mathematics for teaching need to be re-interrogated? My only advice to myself seems to be obvious: to keep reading and listening, and keep an open mind; that is, there are times when my best course of action is to exemplify those habits we want our students to develop.

Discussion

Episode 1 illustrates how Susan's MKT grew because our dialogue was critical – we examined legitimate concerns of teaching – and not because of my telling behaviours as a mentor. Susan's growth is also bound up in her beliefs about teaching based on her past experiences, her current desires as a novice teacher, and her strong ability to do mathematics. Episode 2 illustrates how my MKT grew partially by accident but also because of a personal desire to navigate expert and learner positionings. My growth parallels the way I would like mathematics teachers to grow, and yet I was supposedly already the expert engaged in triggering the professional learning of others. Similarly, episode 3 illustrates how my MKT grew because of my dispositions, first because I wanted to be a learner alongside the teachers and second because I wanted to validate the expertise of teachers. As a mentor, I chose to be silent rather than try to be directive, dialogic or responsive. In terms of MKT, my understandings of MKT evolved, but it is not clear what teachers should know about subitizing. Episode 3 also foregrounds a global uncertainty with my MKT: in what other ways does my MKT need to evolve, and then need to be translated into understandings that can be made available when necessary as a mentor?

From the episodes described above, I see three themes that illuminate notions of mentoring and growth of MKT. First, mentoring models that consider the continuum between entirely responsive to entirely directive dispositions toward a mentee are insufficient because they are silent on the positionings always evident in relationships. I was positioned as expert – by Susan as answer giver – and in response I acted directively. Nevertheless, the episode was still dialogical and critical. When

I sought to position myself as a learner, I found myself learning the same things as Amy did. And being lost in translation with Helen and Cynthia triggered an evolution of my understandings of subitizing. In all three cases, I sought to position myself as validating the expertise of teachers, but in each case I approached this goal differently. The richness and complexity of relationships suggests to me that any model of mentoring will always be insufficient.

Second, it is well established that learning is bound up in who we are. Thus, it is not surprising to bring forward evidence that growth of MKT is inseparable from values, beliefs, dispositions and past experiences. Susan's learning was driven by her personality, current teaching needs, beliefs about teaching and her identity in relation to mathematics. My learning emerged by accident and perhaps by an ability – either trained or personality driven – to notice and reflect (on student learning in episode two, and on competing understandings in episode three). We are all different. This makes the growth of MKT unpredictable; at times random, at times esoteric, at times fundamental.

Third, regardless of how well grounded in practice any analysis of teacher knowledge is, it will always be problematic to fully describe the how, what and when of MKT for every (effective) mathematics teacher. For example, it can easily be argued that every teacher needs a deep, rich, conceptual understanding of mathematical formulas. Is it enough to develop such understandings for some formulas, confident that such understandings will emerge naturally in-situ, as was the case for me with Amy and the area of a circle? My understandings of subitizing are now deep, rich, conceptual, triggered by being lost in translation during a shift in working with pre-service to working with in-service teachers. But I doubt teachers would need the same kind of deep, rich conceptual understanding. I am now better positioned to respond to teachers concerning concepts such as subitizing and the area of a circle, but I still have more to learn. Should I stop working with teachers until my MKT is richer? Clearly no, because I learned by working with teachers. Should Cynthia learn more about subitizing? Is Helen disadvantaged by her understandings of subitizing? Maybe so, maybe not. Should every teacher experience what Amy and I learned about circle area? Maybe not. Dispositions toward mathematics and mathematics learning are just as important because they make possible the kinds of in-situ learning described in this chapter.

Conclusion

This paper considers some of my difficulties in translating theory into practice. Applying theory to practice is not simple; rather, the problematic aspects of theory – in this case, mentoring and MKT – underlie the tensions that emerge in the practice of working collaboratively with teachers. From noticing aspects of the problematic nature of theory and tensions of practice, I have illuminated particular themes, namely, that mentoring models are incomplete, and that growth of MKT is unpredictable and problematic as it is bound up in who we are as teachers and our dispositions toward mathematics and mathematics teaching. These themes lead me to conclude with some reflections concerning the complexities of mentoring and mathematics teacher education, which involve what it means to be a mentor and mathematics teacher educator.

What does it mean to be an effective mentor, in ways that trigger the growth of MKT? The themes of this paper suggest this question is impossible to answer completely. Given that mentoring models must be incomplete, no amount of study or experience can prepare a mentor for the complexities of learning in-situ. I tried to enact the theory, by building trusting, responsive and critical relationships with teachers. In practice, tensions emerged because it was necessary to decide how to enact the theory, such as sometimes telling and sometimes remaining silent. That the growth of MKT is unpredictable, bound up in who we are and our dispositions, suggests that mentors can design opportunities for, but cannot predict professional learning. In the end, I "did" mentoring by being me and attending to what I could learn from the experiences. My work with teachers seemed to work out – there was growth in MKT for me and participating teachers – not just because of my (and others') knowledge of theory, but because of the complexity of practice.

The themes of this paper also suggest mathematics teacher educators, when they work closely with teachers in collaborative environments, must take up challenging roles. Rich descriptions of MKT lead to recommendations for mathematics teacher education curriculum, but I have illustrated difficulties in deciding what should be learned, how it is learned and when. And yet teacher educators must make decisions about content, knowing that dispositions are just as important and probably not teachable. In our relationships with teachers, we are positioned as experts and yet we are still learning. At times, my MKT is inadequate for working with teachers; and yet, it is by working with teachers that my MKT has deepened. When working with teachers, we must navigate roles that can conflict, such as facilitator and purveyor of knowledge. Impossible as it seems, we have to accept these roles, facing the tensions and resisting any desire to simplify the complexity of applying theory to practice.

The reflections above seem to suggest that (mathematics) teacher education is an impossible task. But I am not claiming that we should give up. Nor am I attempting to diminish the work of others. I am only suggesting that maybe teacher educators need to consider shifts in their own dispositions toward teacher professional learning: that we cannot control teacher learning; that professional learning is often unexpected and accidental, and so all we can learn is what we happen to attend to; that anything we learn from books is only theory until it is enacted, and this is the shift into praxis that can enrich our ability to teach or mentor.

References

Ball, D. L., Thames, M. H., & Phelps, G. (2008). Content knowledge for teaching: What makes it special? *Journal of Teacher Education*, *59*(5), 389–407.

Borko, H., Eisenhart, M., Brown, C. A., Underhill, R. G., Jones, D., & Agard, P. C. (1992). Learning to teach hard mathematics: Do novice teachers and their instructors give up too easily? *Journal for Research in Mathematics Education*, *23*(3), 194–222.

Bullough, R. V., Jr., & Draper, R. J. (2004). Making sense of a failed triad: Mentors, university supervisors, and positioning theory. *Journal of Teacher Education*, *55*(5), 407–420.

Burns, M. (1996). *50 Problem-solving lessons: The best from 10 years of math solutions newsletters*. White Plains, NY: Cuisenaire Co.

Clements, D. H. (1999). Subitizing: What is it? Why teach it? *Teaching Children Mathematics*, *5*(7), 400–405.

Darling, L. F. (2001). When conceptions collide: Constructing a community of inquiry for teacher education in British Columbia. *Journal of Education for Teaching*, *27*(1), 7–21.

Dehaene, S., & Cohen, L. (1994). Dissociable mechanisms of subitizing and counting: Neuropsychological evidence from simultanagnosic patients. *Journal of Experimental Psychology: Human Perception and Performance*, *20*(5), 958–975.

Grisham, D. L., Ferguson, J. L., & Brink, B. (2004). Mentoring the mentors: Student teachers' contributions to the middle school classroom. *Mentoring and Tutoring*, *12*(3), 307–319.

Halai, A. (1998). Mentor, mentee, and mathematics: A story of professional development. *Journal of Mathematics Teacher Education*, *1*, 295–315.

Hill, H., & Ball, D. L. (2009). The curious – and crucial – case of mathematical knowledge for teaching. *Kappan*, *91*(2), 68–71.

Jeruchim, J., & Shapiro, P. (1992). *Women, mentors, and success*. New York: Fawcett Columbine.

LeFevre, J., Fast, L., Skwarchuk, S., Smith-Chant, B., Bisanz, J., Kamawar, D., et al. (2010). Pathways to arithmetic: Longitudinal predictors of performance. *Child Development*.

Levin, B., & Rock, T. (2003). The effects of collaborative action research on preservice and experienced teacher partners in professional development schools. *Journal of Teacher Education*, *54*(2), 135–149.

Maynard, J., & Furlong, J. (1993). Learning to teach and models of mentoring. In D. McIntyre, H. Hagger, & M. Wilkin (Eds.), *Mentoring: Perspectives on school-based teacher education* (pp. 69–85). London: Kogan Page.

Meredith, A. (1995). Terry's learning: Some limitations of Shulman's pedagogical content knowledge. *Cambridge Journal of Education*, *25*(2), 175–188.

Morton, M. (2005). Practicing praxis: Mentoring teachers in a low-income school through collaborative action research and transformative pedagogy. *Mentoring and Tutoring*, *13*(1), 53–72.

Orland-Barak, L. (2005). Lost in translation: Mentors learning to participate in competing discourses of practice. *Journal of Teacher Education*, *56*(4), 355–366.

Portner, H. (2008). *Mentoring new teachers* (3rd ed.). Thousand Oaks, CA: Corwin Press.

Shulman, L. S. (1987). Knowledge and teaching: Foundations of the new reform. *Harvard Educational Review*, *57*, 1–22.

Smith, M., Hughes, E., Engle, R., & Stein, M. (2009). Orchestrating discussions. *Mathematics Teaching in the Middle School*, *14*(9), 548–556.

Young, J. R., Bullough, R. V., Jr., Draper, R. J., Smith, L. K., & Erickson, L. B. (2005). Novice teacher growth and personal models of mentoring: Choosing compassion over inquiry. *Mentoring and Tutoring*, *13*(2), 169–188.

Zimmermann, G., Guinee, P., Fulmore, L., & Murray, E. (2009). *Empowering the mentor of the beginning mathematics teacher*. Reston, VA: National Council of Teachers of Mathematics.

Chapter 3
Team-Teaching About Mathematics for All Collaborative Self-Study

Hafdís Guðjónsdóttir and Jónína Vala Kristinsdóttir

Introduction

The diversity of students in Icelandic classrooms is greater than ever. In consequence teachers are currently faced with new challenges to differentiate teaching. The wide spectrum of different students and learning needs requires teachers to adopt flexible approaches to teaching and learning. Furthermore many teachers have discovered that collaborative approaches, such as team-teaching, create more opportunities to enrich pedagogies and respond creatively to all students.

This chapter explains the development of our collaboration as team-teachers of the course *Mathematics for All* which is conducted for graduate teachers at the University of Iceland School of Education. We explain how we used self-study methodology to create, develop and implement a course in team-teaching and mathematics education. Three critical elements define this course: (1) the belief that all children can learn mathematics if they are allowed and supported to construct their own understanding in a community with others, (2) a focus on teaching about team-teaching and the opportunity to experience it in action, and (3) the engagement of teacher learners in the exploration of their own way of learning mathematics. In addition to learning about how children learn mathematics, the teachers develop a clearer understanding of their own way of learning and using mathematics. Questions raised by the teacher learners are important, but equally important are our questions as teacher educators and members of the learning team. Consistent with our beliefs, we encourage the teachers to find their own way to support children's mathematics learning, in a community that promotes learning. Our responsibility is to guide them, and in doing so we provide teacher learners with tools to critically reflect on their practice, to explore children's learning and to respond to their own analyses. Throughout this chapter we refer to university students as teachers, teacher learners or participants, to their students as students or children and to ourselves as teacher educators.

H. Guðjónsdóttir (✉)
School of Education, University of Iceland, Reykjavík, Iceland
e-mail: hafdgud@hi.is

S. Schuck, P. Pereira (eds.), *What Counts in Teaching Mathematics*, Self Study of Teaching and Teacher Education Practices 11, DOI 10.1007/978-94-007-0461-9_3,

All Children Can Learn Mathematics

All through the twentieth century, mathematics and special education developed separately with different disciplines, cultures and professional practices, a dissonance that often contributed to the difficulties experienced by students with disabilities and/or additional learning needs.

The terms "integration" and "inclusion" have been used to describe different levels of including children in school. According to Ainscow (1995) integration makes a limited number of additional arrangements for individual pupils with special educational needs in schools, but the schools do not change much. Inclusion implies more radical changes through which schools restructure themselves to be able to embrace all children. We believe inclusion goes beyond special education, and we support a definition of inclusion that refers to all students and presents inclusive education as a process or an ongoing project where difference is central (Armstrong, Armstrong, & Spandagou, 2010). However, from previous courses we have learned that teachers find it difficult to include all children in the learning community in their mathematics classrooms. Our challenge is to support them in teaching inclusively.

Ainscow (2007) advises teacher educators who embrace diversity and inclusion to learn how to look carefully, to understand the practice as it is carried out in their own classrooms, in their own countries, in order to find starting points to move along the journey. In schools that make progress toward inclusive ways of working, there is a capacity within the school for teachers to learn from one another, to share ideas, and practices, and to spend time talking about how teaching can be improved.

In 1996 Zeichner advised teacher educators to support the belief that all students can succeed if prospective teachers learn scaffolding techniques to bridge students' home cultures and the school culture. He encouraged course work and field experiences in which prospective teachers learn about the languages, cultures and socioeconomic circumstances of the particular students in their classrooms. Also important are learning and teaching strategies that value sense-making and knowledge construction and acknowledge the involvement of parents and other community members in authentic partnerships (Zeichner, 1996).

In mathematics education transformative pedagogy is equally important. Moore (2005) discussed the importance of transformation from theory to practice and concludes that if teachers are expected to teach for diversity and understanding, they need opportunities to develop and enhance their mathematical pedagogical knowledge. It is important for them to experience their own mathematics learning in an environment that reflects the one they are expected to create in their classroom. Teachers are empowered to practice a culturally responsive and socially relevant pedagogy as they begin to look critically at their classroom environment. The practitioner becomes the action researcher, transforming theory into practice and research on that practice.

Learning mathematics with understanding should not only be for students considered to be bright but also be accessible for all students, although all students will not learn the same mathematics to the same degree. According to Marlowe and Page (2005) students with learning difficulties in mathematics are less likely to receive constructivist approaches and more likely to spend time on tasks requiring little more cognitive energy than rote memorization – an approach historically supported by many mathematics teachers and special educators. Hiebert et al. (1997) confirm that many classrooms give students with special needs more instruction on basic skills and less opportunity to develop conceptual understanding. In addition a high percentage of children from poor families, minority groups, and/or who are girls receive more instruction on the mastery of basic skills and less instruction on developing conceptual understanding and learning and how to apply that conceptual understanding to solve authentic problems. The most important role for the teacher is creating a classroom in which all students can reflect on mathematics and communicate their thoughts and actions. Building a community of mathematical practice requires teachers to take the lead in establishing appropriate expectations and norms (Hiebert et al., 1997).

In classroom cultures that promote mathematical learning all students have a voice and are supported to develop their understanding of mathematics through exploring, investigating, discussing, reflecting and drawing conclusions. As well, reflective discussions should support teachers in their teaching:

> The teacher's assessment of a student's conceptual structures does not have to be a blind conjecture. If one starts from the assumption that students generally try to make sense of their experience, it is usually possible to get some idea of how they think. The more experience with learners a teacher has gathered, the better the chance to make an educated guess about what a particular student's thinking might be and to hypothesize what Vygotsky aptly called 'the zone of proximal development'...... It is only after working with a particular student for a considerable time, that the teacher may gain confidence in his or her conceptual portrait of that individual. (Glasersfeld, 1995, p. 187)

Consistent with the belief that teachers need support in transforming theory into practice, we introduce them to the Cognitively Guided Instruction (CGI) program which is focused on: the development of students' mathematics thinking; the instruction that influences that development; teachers' knowledge and beliefs that influence their instructional practices; and the way that teachers' knowledge, beliefs and practices are influenced by their understanding of students' mathematical thinking (Carpenter & Moser, 1984; Carpenter, Fennema, & Franke, 1995; Carpenter, Fennema, Franke, Levi, & Empson, 1999). CGI grew out of inquiry on explicit knowledge about the development of children's mathematical thinking (Carpenter & Moser, 1984) which was used as a context to study teachers' knowledge of students' mathematical thinking (Carpenter, Fennema, Peterson, & Carey, 1988). The rationale was that teachers might use knowledge of students' mathematical thinking in making instructional decisions. In a series of studies (Carpenter, Fennema, Peterson, Chiang, & Franke, 1989; Fennema, Franke, Carpenter, & Carey, 1993; Fennema et al., 1996) it was found that learning to understand the development of children's mathematical thinking could lead to fundamental changes in teachers' beliefs and practices and that these changes were reflected in students' learning.

There was coherence in the change of the teachers' beliefs in the CGI studies and in their way of working with their students. When children began to show increased learning, the teachers continued to implement new methodologies that resulted in improved learning, and so the cycle of developmental change continued. When they found that their students were able to find their own solution strategies when given meaningful problems to work on, they started to rely on the children's thinking. The CGI study provides strong evidence that knowledge of children's thinking is a powerful tool that enables teachers to transform their teaching and change instruction. It also appears that this knowledge is not static and acquired outside the classrooms in workshops, but is dynamic and ever growing, and can probably only be acquired in the context of teaching mathematics (Fennema et al., 1993, 1996).

The CGI project has developed into research on children's algebraic thinking (Carpenter, Franke, & Levi, 2003). Throughout the elementary grades, children are capable of learning powerful unifying ideas of mathematics that are the foundation of both arithmetic and algebra. Learning mathematics with understanding, therefore, should not be reserved for a few mathematically gifted students, it is most critical for students at risk of failing in mathematics to engage in activities that make sense to them.

Another tool we use to support the teacher learners is the Compass Model (Dalvang & Lunde, 2006). The model helps teachers to develop as inclusive teachers who focus on diverse learners' needs and strengths as well as their mathematical learning competencies. According to Dalvang and Lunde the main obstacle in special education is the focus on students' disabilities instead of their abilities. These authors present the Compass Model as a guide for analysis of students' learning conditions and their mathematical competencies. The model assists the teacher in creating a learning community for students based on analysis of the current situation, with a special focus on students' strengths. The students' prerequisites are discussed and the content and design of the current teaching is reconsidered with support from the Danish "KOM" project (Niss & Højgaard Jensen, 2002). The "KOM" project defines important mathematical competencies or proficiencies for pupils to acquire. The Compass Model is built on Niss's and Jensen's definition of mathematical competencies. According to them it is important for all mathematics students to learn to question and answer with mathematics as well as to use its tools and language. We have found that this model helps teachers to build new learning communities for their students.

Processing Team-Teaching

We, the two authors of this chapter, found that our common backgrounds as primary school teachers with inclusive education as a frame for our practice, plus our individual professional knowledge and expertise as a mathematics or special educator prepared us uniquely for this project. Working on our goals as primary school teachers and later as graduate students at different universities, we found it important to have a forum for discussing our work and our studies. Therefore, we met regularly for collaborative reflections on our practice. Continuing the professional discussion, as we became teacher educators, it was natural that we began collaborating between programs in mathematics and special education. During this ongoing time of collaboration we realized that our professional working theory is interrelated, although we at the same time bring in different experience and knowledge from different theoretical fields. Through communication and by participating in a collegial dialogue sharing our professional working theories, we composed a team profile (Dalmau & Guðjónsdóttir, 2002).

The students who take the *Mathematics for All* course are teachers (pre-, primary and secondary) and social educators. Some have a strong mathematical background and have taught mathematics for a long time; others have little experience of formal mathematics teaching. The course is mainly organized through distance learning, although participants also meet on campus for five full days. Internet-based software is used for teaching and learning in between campus meetings. Analysis of our original courses revealed that participants found it important to understand the impact of learning disabilities on mathematics education. They also recognized the importance of understanding and challenging their beliefs, knowledge and practices. We learned that some participants felt uncertain about mathematics and were not confident in teaching the subject (Guðjónsdóttir & Kristinsdóttir, 2006). These results reflect participants' diverse backgrounds.

The predominant emphasis, in the first years we taught the course, was on the special education in mathematics and the connected areas of learning disabilities, identification/diagnosis of problems and errors, and teaching algorithms. As the course developed we worked with teacher learners to shift the emphasis *from* diagnosing children's weaknesses *to* exploring how children learn mathematics and to building teaching on students' strengths rather than their weakness. As time has passed, we have learned more about the teachers' dilemmas related to their beliefs and attitudes, their professional understanding and the ways they themselves learned mathematics when they were children (Guðjónsdóttir & Kristinsdóttir, 2006, 2007). Teachers' conceptions of mathematics, teaching and learning, gained during their early educational experiences, make it difficult for them to act in new ways that foster a different quality of mathematics learning for students (Crawford & Adler, 1996). It is important for teachers to explore and experiment with mathematics and engage in discussions about their own mathematical thinking as well as learning about the development of children's mathematical learning (Bredcamp, 2004).

Team-teaching is an approach where two or more teachers take joint responsibility for course content, activities and assessment. All teachers belonging to the team are actively involved in the lesson and share the teaching in all ways (Bacharach, Heck, & Dahlberg, 2008). It requires a commitment to collaborative teaching based on school cultures that encourage professionals to work together (Barth, 2006). Through the years we have been jointly responsible for the course content, presentations and grading. We have interacted in front of the class and discussed specific problems from different points of view (Schafer, 2000–2001; Wallace, 2007). As noted above, our varied experience created an ideal climate in which to build together, a community where differences are respected, and can be discussed and used to support our understanding and learning. This is supported by Russell (1997) who claimed that the way teacher educators teach demonstrates the potential for how student teachers can teach. Crow and Smith (2003) report that their team-teaching had a positive impact on both their students and themselves. Their students commented that the different background the professors brought to the course expanded their learning options, changed how they worked and gave them an example of what they were learning about. Our experience is similar. By ongoing collaborative reflections on our work, we were able to revise the course with the intention of building a community where our students interact, share their ideas and draw conclusions from their explorations in mathematics and in their teaching. It is complicated to teach about teaching, but the constant reviewing and reconstruction which keeps alive what we do is what makes the job so interesting (Kroll, 2007).

Collaborative Self-study

This research is a self-study of our teaching as we develop, reconstruct and team-teach a graduate course: *Mathematics for All*. Self-study positions the teacher both as an inquirer and a learner (Samaras & Freese, 2006), as the purpose of the self-study inquiry is to better understand the practice, the renewal of programs, and the personal–professional development (Kosnik, Beck, Freese, & Samaras, 2005). By collecting, interpreting and evaluating evidence from different sources, educators create opportunities to achieve their aim. Based on both personal and practical experience of teacher educators, self-study research leads to collaborative, questioning, dialogic and action-oriented processes (Bodone, Guðjónsdóttir, & Dalmau, 2004). In other words, it is a methodology of studying professional practice settings that allows teachers and teacher educators to maintain focus on both their teaching and their students' learning (LaBoskey, 2004; Loughran, 2004).

Self-study does not exist without some form of collaboration (Bodone et al., 2004). In our study, collaboration was grounded in the study from the beginning, giving us opportunities to reflect critically on the development of our teaching and the course we are studying. The process of action-reflection-learning-action that is at the heart of collaborative self-study research provided three significant opportunities for the renewal of the course:

- collection of data about teaching and learning;
- deeper understanding of how we can teach teachers to analyze their own practice and their students' learning; and
- the review and improvement of our team-teaching and research collaboration.

While the focus of this study is on ourselves as professionals, and on our roles in teaching and developing the course, we also believe that we cannot leave the teacher learners out of the research or out of the story we report. Thus we present vignettes of the interaction with them.

We have conducted continuous self-study research since the establishment of the mathematics course 7 years ago. Data were gathered from multiple sources, including all the material from the courses, student projects, class discussions and documentation of our teaching, discussions and reflections. The research process formed a spiral where data collection, data analysis and interpretation, and teaching are intertwined as we plan our actions for the next class. The case and commentary by the teacher learners was analyzed to understand what and how the teachers learnt about their students' learning. From our analyses and interpretation we decided what and how to teach next, or what reading to bring into the class. To better understand the impact of the course on teachers' work, we interviewed them, and observed their classrooms.

To make the data upon which we base our study accessible and clear, we offer vignettes from our classrooms or teaching that are drawn from field notes, teachers' work and our collaborative reflections (Loughran, 2007). The vignettes are used to bring to life certain situations or episodes from our teaching and collaboration, and are used both as a method and a form of analysis (Loughran, Berry, & Tudball, 2005). The double collaboration, teaching and researching, offers opportunities to see our teaching and the teacher learners' work from different perspectives. By team-teaching we experience what happens in the classroom, and because of our different roles and theoretical backgrounds, it gives us different perspectives. Thus we are able to respond in a way that enhances our teaching and deepens our research.

Collaborative Practice

We begin each teaching term by emphasizing the teacher learners' diversity and their many different ways of learning to solve mathematical problems. We also encourage teachers' own investigations of mathematics and exploration of problems that has the potential to promote mathematical activity and thinking and stimulate collaboration with meaningful discussions and shared thinking (Jaworski, 2007). The spiral of building confidence; manipulating; getting sense of; capturing in pictures, words and symbols; providing fodder for further manipulation etc, as explained by Mason (1999), is used as a guiding tool in our explorations with the mathematics. We have found these guidelines helpful because the teachers are not used to exploring mathematics.

The teachers work in groups, and we encourage them to write down their thinking and the steps they are taking to find the solution. We have created the following questions to scaffold their thinking and reflection.

- What was my first thought? What did I do next?
- What new ideas emerged in discussions with students in the group?
- How did I contribute to the discussions?
- Did my ideas help develop the discussions?
- Did I learn something from other students' ideas and explanations?

Finally the groups report their solution/strategies and we discuss different approaches. We participate in the problem-solving process by joining the group discussions and guiding the reporting.

A vignette from one of our workshops illustrates how this worked:

> We posed the following problem:
>
> *Siggi baked cookies for his 12 grandchildren and put them into bags. Some of the cookies were big and others small. He decided that the big ones would go 7 in a bag and the small ones 8 in a bag. He packed 91 cookies into the 12 bags.*
>
> *How many bags contained big cookies and how many contained small ones?*
>
> Jónína was in charge of the lesson and Hafdís joined her. A group of secondary teachers was discussing different ways to solve the problem and found that the most efficient way would be to make two equations with two unknown variables. A group of preschool teachers was listening to their discussions and seemed frustrated hearing the mathematical language they used. Jónína discussed the problem with them and they admitted that they didn't have a clue how to solve it. She then asked them if it would help them to draw a picture of the situation but they still seemed confused. She suggested that they should try to draw a picture of the situation and then went to find cubes to count. Hafdís was aware of the confusion in the group and continued the discussion while Jónína was away. When Jónína came back Hafdís felt that the teachers were so frustrated that they most of all wanted to leave the room. But Jónína was standing at the door with a tray of 100 cubes which she put on their table so there was no way of escaping. She asked them if they could use the cubes to solve the problem. They hesitated for a while and then gradually began by first taking nine cubes away and then making 12 groups out of the rest.
>
> Jónína asked the preschool teachers if they were willing to share their experience. They hesitated and then started by apologizing for solving the problems like young children. This gave us the opportunity to discuss different ways to solve problems and the diverse solution strategies the teachers used. Both of us participated in the discussion as well as the other teachers. We were happy to find that the secondary teachers supported the preschool teachers by admitting that the preschool teachers' way of solving the problem helped them to realize what their students might gain from having mathematical models available in mathematics lessons.
>
> We discussed the situation that arose in our classroom with the teachers. Do similar situations arise in their classrooms? Can they learn from this experience to deal with teaching in diverse classrooms? Do grouping methods affect the learning in the classroom? We did not divide the teachers into groups so those who knew each other were working together. Would we have had a different experience if we had grouped them differently?
>
> We also discussed the mathematics that they were exploring and how they dealt differently with the problem depending on their mathematical background. Can children learn different things by dealing with a problem like this? Is this a valid problem for their students? What would they learn from solving it? Our interplay during this session was natural and we supported each other constantly in the discussion where our strengths in different fields supported us in the discussion.

This interplay and mutual support helped us create an interactive learning environment. By walking around and participating in the small group discussions, we managed to gain knowledge of the mathematical situation in the whole group, use it for the discussion and point out that it is natural that students in every group think about and solve problems in different ways.

To better understand how our teaching impacted our students, we asked them to discuss and record their reflections on their learning and their perspectives. Their solution strategies differed profoundly, and we discussed how their understanding of the problem reflected their way of learning mathematics. We assessed and discussed the way the teacher learners solved mathematical problems, and from there we related their personal experience to their experience with the children. We supported them to connect to children's learning and how children might understand the problem (Guðjónsdóttir & Kristinsdóttir, 2007).

We have learned that critical reflections on own thinking and work can be a challenge and therefore we continually create and develop different strategies and tasks for scaffolding the teacher learners to understand how their students learn mathematics. One such is the case and commentary analysis based on the praxis inquiry protocol (Kruger & Cherednichenko, 2006). The process consists of five iterative stages: case writing and four dimensions of praxis inquiry.

Case writing: Cases are professional stories that teachers write in order to stimulate their inquiry and analysis on the real challenges and dilemmas of their practices. The four dimensions of the protocol are used as a scaffold to analyze the case.

Practice described: The participant describes the persons in the case and their social situation in sufficient detail to enable them to better understand the action.

Practice explained: The participant, in describing practice, has adopted an explicit discourse or discourses for interpreting the action. As they interpret the practice, they ask themselves what professional explanations they can find to assist them to understand and explain what is happening.

Practice theorized: Participants construct their personal theory of the practice described. At this stage they relate theory and professional working and ground their practice in theory and ask questions like: As I incorporate these understandings who am I becoming as a professional? What are my significant professional practices, beliefs and theories?

Practice changed: Theorized practice presents practitioners with opportunities to propose and trial new practices. At this last step, practitioners make decisions or conclusions, and develop and improve their practice.

This protocol and the systematic way of reflecting critically on practice and relating it to theory helped teacher learners to see their practice from different perspectives and encouraged them to explore research findings and to use their conclusion to develop their practice. They adapted this process to their work as they analyzed their practice, their students' understanding and learning, and figured out how to respond in a professional way.

In her response in the web-based log Björk shared her reflections from the course with us:

> What I found interesting was when we solved the problem in different ways. Then I thought: If we can solve problems in different ways then the children must be able to do it and even in more different ways than we do. They find a strategy that they understand and can even explain how they did it. Their understanding develops and if they invent something themselves they will remember it. Then they will use this, even without knowing it, in their lives (learning) and get used to investigate and explore 'hidden' sides of something. They will be more open, because their understanding was respected.

Björk related her experience to her own teaching and the capacities and interests of children she works with. She observed that children are more likely to understand and use strategies they have invented themselves.

In her response Linda demonstrated her awareness of how necessary it is to give students the opportunity to do things their way.

> . . . I have learned so much and the greatest experience was to discover that problems can be solved in incredibly many ways; and how children use different ways and all kinds of hands on materials to solve them. . . . I'm a swimming trainer for handicapped people and I have always believed that it doesn't really matter how you swim. It is much more important to be able to swim. . . . The goal is that the students can use their capability to swim to escape from drowning and as a tool for training their body and for very few as athletes. In the same way we want everyone to be able to solve mathematical problems, in their own way.

The web-based log is one of the tools we use for encouraging the teacher learners to reflect on their learning. Discussing their work as well as the course readings helped them to connect the theories and their experience as learners and teachers. They lead the discussions on the web and we, the teacher educators, support them by participating in their discussions.

Professional Development

To learn about the impact the course has had on the teacher learners, we have interviewed some of them and observed their classrooms.

Anna teaches at a school where some of her colleagues have worked at developing their mathematics teaching for 15 years, and she feels that she is becoming a part of that development as she builds on her professional development to change her practice. She is aware of how what she learned in the course has affected her teaching.

> What I learned at the course has helped me to plan and structure my teaching. I'm more confident in relying on that children who have difficulties in mathematics can construct their own knowledge. We often believe that they cannot learn with understanding. What they need to do is to practise and learn words that trigger them what to do. I don't believe that any more. I also feel that I'm much stronger in analyzing the children's solution strategies and understanding of the problems.

Helga also finds the support at the school important. The quality of the school as a workplace, shared practice, student-centred values and teacher support, are important dimensions of effective professional development for teachers. She expresses how her teaching has changed.

> I find it important to teach children according to their abilities and understanding. I use my intuition to see and hear what they are thinking and how they learn. I didn't do that so much before. I didn't give myself time to do that and didn't find it as important as I do now. The learning situation is also more lively and supportive. I have learned that it matters. But what really triggers me in my work is my interest and enthusiasm to work and learn with children not the least when I witness their victories.

Stina finds her situation at work challenging. She would like to bring more of what she is learning into her practice but is not receiving the support she needs or agreement from other teachers. She realizes that it is not easy to bring what she learned during the course into her practice.

> I am aware of my education and would like to teach mathematics in a different way but I feel that my situation doesn't allow it.

Teachers need support to develop their practice both within their own schools and by participating in professional courses. Teachers' professional development involves making explicit the beliefs and values that underlie their actions and practices and contribute to their learning.

Discussion

Our self-study and inquiry into the development of the course supported us in reflecting together on our teaching and also in reflecting and responding to each other in our teaching (Schön, 1987). Our interplay during the sessions is a result of long-time collaboration and the trust that we have built by reflecting together on pedagogy. Traditionally, special education focuses on students' deficiencies. In our course we explore students' abilities in mathematics. Traditions and knowledge from two disciplines are brought together to create a vision on teaching all children mathematics.

Jaworski (2006) discusses the way inquiry can become a tool for critical awareness. "In communities of inquiry, we all engage with inquiry as a tool to develop meta-knowing, a form of critical awareness that manifests itself in inquiry as a way of being" (Jaworski, 2006, p. 208). It was critical that we were both present in a dual role, as a teacher educator and a researcher. The team-teaching gave us the opportunity to change roles back and forth from teaching to researching and by collecting data, analyzing and interpreting them, to combine teaching and researching and walk the talk (Samaras, 2010).

Our analysis and interpretations have influenced ongoing development of the course. They emphasize children's mathematical development and understanding as well as teachers' capacity to evaluate and promote students' learning through analysis of their engagement in authentic mathematical problems. We move from special education's traditional focus on diagnosing and working on deficiencies toward more authentic assessment of mathematical thinking. We observe how participants solve their problems and have them explain their thinking and their work. The teacher learners benefited from the growth of their content knowledge of mathematics as well as by experiencing a model for teaching that focused on the students' strengths and interests and team-teaching.

All our work, our pedagogy and the trust we have managed to build enabled us to collaborate during our teaching. The dialogue and the constructive feedback have enriched the classroom environment. Reporting on our study, presenting and writing about our research, enhances our understanding of how our collaboration and team-teaching has grown to develop a community of inquiry where we reflect on our work together. It has also affected the learning community that we have developed along with the teacher learners. When the teachers explored mathematics their confidence in solving problems increased. Additionally, as they learned about their differences in problem solving and understanding mathematics, their understanding of how children use diverse ways to solve mathematical problems expands. Problem-solving activities may help teachers to experience and discuss difficulties similar to those met by students in class and understand the importance of evaluating the process in mathematical activities (Boero, Dapueto, & Parenti, 1996; Crespo & Sinclair, 2008).

The teachers' responses and their discussions, both in face-to-face sessions and on the web, imply that they gradually become a part of our community of inquiry, as we urge them to participate seriously in discussions that help them reflect on their ways of learning and teaching. Using the case and commentary praxis inquiry protocol helps them to reflect in a critical way and to relate practice and theory as they develop their practice.

While discussing our learning and reflecting on our experiences during the course, we often speculate why we find it natural to collaborate. We think about our philosophies of education and how they relate to the collaboration. We are both educated as classroom teachers, but Hafdís later specialized in special education and Jónína in general pedagogy and mathematics education. We have both worked for a long time as primary school teachers though in different schools and within different school cultures. At the same time we are both interested in mathematics and inclusive education and have studied and taught together at workshops and university courses in these areas.

Our team comprises instructors from different academic fields. It is therefore important to notice that we bring into our teaching different strengths; however, we have an insight into each other's interests and specialties also. We bring a variety of perspectives to the subject under consideration (Schafer, 2000–2001). Mathematics education and special education have developed from different disciplines, and the knowledge we bring with us makes the team stronger than the individuals that make up the team. As we prepare our course we each bring in different knowledge but at the same time our pedagogy and way of teaching overlaps. According to Pugach (2005) perspectives and attitudes of teachers involved in inclusive education are more positive about diverse groups of students than those who have not had that opportunity.

Conclusions

Our collaboration has helped us see from a broad range of views how our students are learning, and in so doing we believe that we have managed to respond to them in a more professional way. By discussing our responses to them and helping each other understand their learning, we have opened up a forum and encouraged them to critically reflect on their classroom practice in the light of research (Cockburn, 2008). Giving the teacher learners access to research on children's mathematical development and understanding, their capacity to evaluate and promote students' learning through analysis of their engagement in authentic mathematical problems has been enhanced.

According to Loughran (2007) self-study can and should make a major contribution to improving the quality of teacher education. The close relationship between teaching and researching is a motivation for self-study because knowledge development can then be closely tied to one's own teaching, the results of which can then shape the manner in which subsequent practice is conducted. By writing about our research and our collaboration, we believe that we can make the new knowledge meaningful to others.

What we have found important is to:

1. "Walk the talk" and team-teach with members that bring in diverse knowledge and experience.
2. Participate in the inquiry along with the teacher learners. Reflect together with them and discuss our role in the modeling practice.
3. Begin with participants' knowledge, experience and questions, both as teachers and learners.
4. Always emphasize students' strength and ability.

The transformation from theory to practice does not proceed automatically. Teacher educators need to help create learning communities for teachers and should be responsible for supporting them in teaching mathematics in inclusive schools.

References

Ainscow, M. (1995). Education for all: Making it happen. *Support for Learning*, *10*(4), 147–154.

Ainscow, M. (2007). Teacher development in responding to student diversity: The way ahead. In P. A. Bartolo, A. M. Lous, & T. Hofsäss (Eds.), *Responding to student diversity. Teacher education and classroom practice. Proceedings of the International Conference on Teacher Education for Responding to Student Diversity* (pp. 1–22). Malta: Faculty of Education, University of Malta.

Armstrong, A. C., Armstrong, D., & Spandagou, I. (2010). *Inclusive education: International policy & practice*. London: SAGE.

Bacharach, N., Heck, T. W., & Dahlberg, K. (2008). Co-teaching in higher education. *Journal of College Teaching & Learning*, *5*(3), 9–16.

Barth, R. S. (2006). Improving the relationships within the schoolhouse. *Educational Leadership*, *63*(6), 8–13.

Bodone, F., Guðjónsdóttir, H., & Dalmau, M. C. (2004). Revisioning and recreating practice: Collaboration in self-study. In J. J. Loughran, M. L. E. Hamilton, V. LaBoskey, & T. Russell (Eds.), *International handbook of self-study of teaching and teacher education practices* (pp. 743–784). Boston: Kluwer Academic Publishers.

Boero, P., Dapueto, C., & Parenti, L. (1996). Didactics of mathematics and the professional knowledge of teachers. In A. J. Bishop (Ed.), *International handbook of mathematics education* (pp. 1097–1121). Dordrecht and London: Kluwer Academic Publishers.

Bredcamp, S. (2004). Standards for preschool and kindergarten mathematics education. In D. H. Clements & J. Samara (Eds.), *Engaging young children in mathematics* (pp. 77–82). Mahwah, NJ: Lawrence Erlbaum Associates.

Carpenter, T., Fenema, E., Peterson, P. L., & Carey, C. A. (1988). Teachers' pedagogical content knowledge of students' problem solving in elementary arithmetic. *Journal for Research in Mathematics Education, 19*, 385–401.

Carpenter, T., Fennema, E., & Franke, M. L. (1995). *Children's thinking about whole numbers*. Madison, WI: University of Wisconsin: Wisconsin Center for Educational Research.

Carpenter, T., Fennema, E., Franke, M. L., Levi, L., & Empson, S. B. (1999). *Childrenś mathematics: Cognitively guided instruction*. Portsmouth, NH: Heineman.

Carpenter, T., Fennema, E., Peterson, P. L., Chiang, C. P., & Franke, M. L. (1989). Using knowledge of childrenś mathematics thinking in classroom teaching: An experimental study. *American Educational Research Journal, 26*(4), 499–531.

Carpenter, T. P., Franke, M. L., & Levi, L. (2003). *Thinking mathematically. Integrating arithmetic and algebra in elementary school*. Portsmouth, NH: Heineman.

Carpenter, T., & Moser, J. M. (1984). The acquisition of addition and subtraction concepts in grades one through three. *Journal for Research in Mathematics Education, 15*(3), 179–202.

Cockburn, A. D. (2008). How can research be used to inform and improve mathematics teaching practice? *Journal of Mathematics Teacher Education, 11*(5), 343–347.

Crawford, K., & Adler, J. (1996). Teachers as researchers in mathematics education. In A. J. Bishop (Ed.), *International handbook of mathematics education* (pp. 1187–1205). Dordrecht and London: Kluwer Academic Publishers.

Crespo, S., & Sinclair, N. (2008). What makes a problem mathematically interesting? Inviting prospective teachers to pose better problems. *Journal of Mathematics Teacher Education, 11*(5), 395–415.

Crow, J., & Smith, L. (2003). Using co-teaching as a means of facilitating interprofessional collaboration in health and social care. *Journal of Interprofessional Care, 17*(1), 45–55.

Dalmau, M. C., & Guðjónsdóttir, H. (2002). Framing professional discourse with teachers: Professional working theory. In J. Loughran & T. Russell (Eds.), *Improving teacher education practices through self-study* (pp. 102–129). London and New York: Routledge/Falmer.

Dalvang, T., & Lunde, O. (2006). Med kompass mot mestring et didaktisk perspektiv på matematikvansker. *Nordic Studies in Mathematics Education, 11*(4), 37–64.

Fennema, E., Carpenter, T. P., Franke, M. L., & Carey, D. A. (1993). Using children's mathematical knowledge in instruction. *American Educational Research Journal, 30*, 555–583.

Fennema, E., Carpenter, T., Franke, M. L., Levi, L., Jacobs, V. R., & Empson, S. B. (1996). A longitudinal study of learning to use children's thinking in mathematics instruction. *Journal for research in mathematics education, 27*, 403–434.

Glasersfeld, E. von. (1995). *Radical constructivism. A way of knowing and learning*. London: RoutledgeFalmer.

Guðjónsdóttir, H., & Kristinsdóttir, J. V. (2006). Teaching all children mathematics: How self-study made a difference. In L. M. Fitzgerald & D. L. Tidwell (Eds.), *Self-study and diversity* (pp. 195–211). Rotterdam: Sense Publishers.

Guðjónsdóttir, H., & Kristinsdóttir, J. V. (2007). Preparing teachers to teach all children mathematics. In P. A. Bartolo, A. M. Lous, & T. Hofsäss (Eds.), *Responding to student diversity: Teacher education and classroom practice. Proceedings of the International Conference on Teacher Education for Responding to Student Diversity* (pp. 44–61). Malta: Faculty of Education, University of Malta.

Hiebert, J., Carpenter, T. P., Fennema, E., Fuson, K. C., Wearne, D., Murray, H., et al. (1997). *Making Sense. Teaching and learning mathematics with understanding*. Portsmouth, NH: Heinemann.

Jaworski, B. (2006). Theory and practice in mathematics teaching development: Critical inquiry as a mode of learning in teaching. *Journal of Mathematics Teacher Education, 9*, 187–211.

Jaworski, B. (2007). Introducing LCM – learning communities in mathematics. In B. Jaworski, A. B. Fuglestad, R. Bjuland, T. Breiteig, S. Goodchild, & B. Grevholm (Eds.), *Learning communities in mathematics* (pp. 13–25). Norway: Caspar Forlag.

Kosnik, C., Beck, C., Freese, A., & Samaras, A. (2005). *Making a difference in teacher education through self-study: Studies of personal, professional and program renewal*. Dordrecht: Springer.

Kroll, L.. (2007). Central nature of theory in my practice. In T. Russell & J. Loughran (Eds.), *Enacting a pedagogy of teacher education: Values, relationships and practices* (pp. 95–105). London and New York: Routledge Taylor & Francis Group.

Kruger, T., & Cherednichenko, B. (2006). Social justice and teacher education: Re-defining the curriculum. *The International Journal of Learning, 12*, 1–8.

LaBoskey, V. K. (2004). The methodology of self-study and its theoretical underpinnings. In J. J. Loughran, M. L. Hamilton, V. K. LaBoskey, & T. Russell (Eds.), *International handbook of self-study of teacher education practices* (pp. 817–870). Dordrecht: Kluwer.

Loughran, J. (2004). A history and context of self-study of teaching and teacher education practices. In J. J. Loughran, M. L. Hamilton, V. K. LaBoskey, & T. Russell (Eds.), *International handbook of self-study of teacher education practices* (pp. 7–40). Dordrecht: Kluwer.

Loughran, J. (January/February 2007). Researching teacher education practices: Responding to the challenges, demands, and expectations of self-study. *Journal of Teacher Education, 58*(1), 12–20.

Loughran, J., Berry, A., & Tudball, L. (2005). Learning about teaching. In C. Kosnik, C. Beck, A. Freese, & A. Samaras (Eds.), *Making a difference in teacher education through self-study: Studies of personal, professional and program renewal* (pp. 203–226). Dordrecht: Springer.

Marlowe, B. A., & Page, M. L. (2005). *Creating and sustaining the constructivist classroom*. Heatherton, Victoria Australia: Hawker Brownlow Education.

Mason, J. (1999). *Learning and doing mathematics*. York: QED.

Moore, J. (2005). Transformative mathematics pedagogy: From theory to practice, research, and beyond. In A. J. Rodriguez & R. S. Kitchen (Eds.), *Preparing mathematics and science teachers for diverse classrooms*. Mahaw, NJ: Lawrence Erlbaum.

Niss, M., & Højgaard Jensen, T. (Red.). (2002). *Kompetenser og matamatiklæring – ideer og inspirasjon til udvikling af matematikundervisning i Danmark*. København: Undervisningsministeriets forlag.

Pugach, M. C. (2005). Research on preparing general education teachers to work with students with disabilities. In M. Cochran-Smith & K. M. Zeichner (Eds.), *Studying teacher education: The report of AERA panel on research and teacher education*. Mahwah, NJ: Lawrence Erlbaum associates.

Russell, T. (1997). Teaching teachers: How I teach IS the message. In J. Loughran & T. Russell (Eds.), *Teaching about teaching: Purpose, passion and pedagogy in teacher education* (pp. 32–47). London: Falmer Press.

Samaras, A. P. (2010). *Self-study teacher research: Improving your practice through collaborative inquiry*. Thousand Oaks, CA: Sage.

Samaras, A., & Freese, A. (2006). *Self-study of teaching practices*. New York: Peter Lang Primer.

Schafer, I. (2000–2001). *Team teaching: Education for the future*. Retrieved March 14, 2008, from http://www.usao.edu/~facshaferi/teamteaching.htm.

Schön, D. A. (1987). *Educating the reflective practitioner*. Oxford: Jossey-Bass.

Wallace, J. J. (2007). Effects of interdisciplinary teaching team configuration upon the social bonding of middle school students [Electronic Version]. *Research in Middle Level Education*, 5. © 2007 National Middle School Association?

Zeichner, K. (1996). Educating teachers for cultural diversity. In K. Zeichner, S. Melnick, & M. L. Gomes (Eds.), *Currents of reforms in preservice teacher education* (pp. 133–151). New York: Teachers College Press.

Chapter 4
Growing Possibilities: Designing Mathematical and Pedagogical Problems Using Variation

Cynthia Nicol

I am a beginning gardener. My yard is large, much of it overgrown, and each year I take on a small section to transform from thorns and brush to cultivated plants and flowers. For some this might be a simple task, but for me it is complex. Trying to figure out what to plant where, and trying to anticipate what it will look like in 5 or 10 years is a challenge. I have many gardening books that list various plants and shrubs alphabetically, or by plant type, or by the amount of sun or shade the plants prefer. My gardening friends also provide ideas and possibilities for what to plant. I am becoming familiar with certain varieties of plants and I can quickly locate descriptions in my resources. However what I have not yet developed is a sense for selecting plants that can live and grow together in specific places in my garden. I find myself wanting someone to just tell me which plants to purchase and where they should be planted. Not unlike my teacher candidates, living in frustration due to lack of experience and skill, but with the desire to do a good job, I want to be told just how to do it and how to do it well.

Teaching, like gardening, requires more than access to lists of references, resources, problems and textbooks. It requires more than a list of good questions to pose to students to explore their mathematical thinking and more than a list of how to respond. These are useful resources but unless teacher candidates have the experience, skill and confidence to know when and how to use them they remain inert and disconnected from practice. However, many teacher candidates have experienced learning mathematics as a series of rules for solving uncomplicated problems and see teaching mathematics as explaining procedures for solving such problems and assigning homework to practice those skills. They come to mathematics methods courses ready to learn the most efficient and effective ways to show students how to use procedures. They are ready to be told how to do it and how to do it well.

When introduced to alternatives to ways they previously learned mathematics, many teacher candidates worry that they do not yet have the necessary mathematical

---

C. Nicol (✉)
Department of Curriculum and Pedagogy, University of British Columbia, Vancouver, BC, Canada
e-mail: cynthia.nicol@ubc.ca

S. Schuck, P. Pereira (eds.), *What Counts in Teaching Mathematics*, Self Study of Teaching and Teacher Education Practices 11, DOI 10.1007/978-94-007-0461-9_4,

knowledge or teaching experience in order to teach differently. Although many are excited to learn of teaching approaches that lead to understanding and sense-making, they are overwhelmed by the task of unlearning their previous mathematical ideas in order to consider teaching differently. My goal is not to convince teacher candidates of one way to teach mathematics, but instead to increase their awareness of difference, that is, that there are different ways to consider teaching mathematics. Deciding why, when and how to draw upon one strategy over another is part of what makes teaching an intellectual and moral endeavour. Thus a further goal for me is helping teacher candidates develop a stance of inquiry or as Featherstone (2007) states, "a questioning stance" related to teaching and learning mathematics. Developing such a stance requires that I too examine my practice alongside my students and my colleagues.

This chapter draws upon my experience as a mathematics teacher educator and my work developing opportunities to practice inquiry with my colleagues and teacher candidates. The chapter focuses on my work at two levels. At the first level I examine my role as an instructor of a problem-based learning cohort in a 12-month post-baccalaureate elementary teacher education program. At the second level I examine the process of participating in a collaborative self-study group that explores ways to design tasks for teacher candidates to provide opportunities for meaningful learning of mathematics. In the following section I explore the nature of collaborative self-study before introducing the theoretical considerations that framed my work.

Collaborative Self-study

There are multiple research methods for engaging in self-study of practice (Tidwell, Heston, & Fitzgerald, 2009). Although collaborative self-study appears at first to be a contradiction, collaboration is often conceptualized as an essential aspect of self-study research (Loughran & Northfield, 1996). Coia and Taylor (2009) extend this idea with their collaborative self-study method that recognizes not only the individual or self but also the self in relation to others. They noted "we can research ourselves only within the context of others" (p. 7). Bodone, Guðjónsdóttir, and Dalmau (2004) concur that collaboration could be the focus of inquiry in a self-study, it could be a method of working with another person, team or group, or it could involve a dialogue of the researcher with literature, texts or past experiences. With such a broad definition of collaboration it is not surprising that Bodone and colleagues "found no examples of self-study that did not involve some form of collaboration" (p. 771).

Although all self-study may contain some form of collaboration, I have found it useful to articulate three models of collaborative self-study in the context of mathematics teacher education (Nicol, Novakowski, Ghaleb, & Beastro, 2010). One model is defined as *cross-disciplinary* or *multi-party collaborative self-study* (Bodone et al., 2004). Here teacher educators meet regularly to share their insights, questions and knowledge of their practices. Members need not share the same content area or teaching assignment but instead meet to examine a shared interest within their respective contexts. A second model I term *concept-based collaborative self-study*. In this model, teacher educators select a mathematical topic and meet to discuss student learning in relation to this topic and the challenges of teaching it. They move to their respective classrooms to investigate the topic and return to the group to share their experiences of self-study (e.g., Marton & Tsui, 2004; Stigler & Hiebert, 1999). A third model involves teacher educators working together, typically in pairs or small groups, within the same class (e.g., Loughran & Northfield, 1996; Roth & Tobin, 2005). As this model involves collaboration with one or more critical friends, it is defined as *critical collaborative self-study* (Nicol et al., in press).

Certainly there are hybrid versions of these models of collaborative self-study work. "New forms of collaboration are therefore necessary", state Bodone and colleagues, "if educators are to access the partners, viewpoints and capacities that will be necessary if self-study is to achieve its goals" of re-imagining teacher education (p. 775). As a teacher educator I have, over the years, practiced collaborative self-study in multiple contexts. My current work is located in the spaces between concept-based collaborative self-study and critical collaborative self-study research.

I am a member of a collaborative self-study group called the Mathematics Teacher Education Collaborative (MTEC). This group includes elementary and secondary level teacher educators, graduate students, practising teachers and sessional instructors interested in both improving their practice as mathematics teacher educators and understanding teacher candidates' experiences and learning. We meet approximately every 4–6 weeks as a group to share ideas, questions, and insights of our practice. For this chapter I focus on my experiences within the group and our efforts to collaboratively study the nature of mathematics teacher education in order to improve the opportunities we provide our students.

Mathematics Teacher Education Collaborative

A collective goal of MTEC is to examine how our teacher candidates learn three central practices of mathematics teaching: questioning, listening and responding (Nicol, 1999). Questioning involves selecting and posing mathematical problems to students. The practice of listening involves interpreting students' mathematical work, while responding entails acting in response to students' mathematical ideas. MTEC brought us together to explore and examine how we as teacher educators could develop pedagogical problems for teacher candidates (TC) to gain more experience with these three practices of questioning, listening and responding. Thus, at a meta-level MTEC provides a space to examine our own questioning, listening and responding as mathematics teacher educators. MTEC members teach a variety of courses but most teach mathematics methods to elementary TCs. I teach in a problem-based learning cohort and am responsible for teaching mathematics curriculum and pedagogy to the elementary TCs in this cohort. I am the only member of MTEC teaching mathematics in such a context. Like other cohorts within our program, the Problem-Based Learning (PBL) cohort is part of a 12-month post-baccalaureate teacher education program. However, it differs in its interdisciplinary approach and focus on learning through cases or problems of teaching. The cases are organized around issues and dilemmas, and it is my role as mathematics teacher educator to help TCs examine and explore the issues from a mathematical and pedagogical perspective.

My colleagues and I have over the years used video cases in our methods courses as a way to provide opportunities for TCs to examine mathematics practice. Video cases of mathematics teaching and interviews of children reasoning through mathematics problems such as those developed by Carpenter and colleagues as part of the Cognitively Guided Instruction (CGI) research project (1999, 2003) formed the basis of much of my work with TCs. Video cases along with written excerpts of students' school work became artifacts for discussion and analysis. Through these artifacts our central MTEC goal of preparing TCs to question, listen and respond to students' mathematical thinking was developed. Similar to the mathematics teacher education practices articulated by Roth McDuffie, Drake, and Herbel-Eisenmann (2008), MTEC members saw the value of teaching TCs through making sense of children's mathematical thinking.

In my problem-based learning cohort I used children's thinking as a place to challenge and extend TCs' assumptions on the possible mathematics problems that could be posed to students, on how students can solve challenging problems and communicate their thinking, and on how teachers might respond. Generally, TCs began by solving the mathematics problem themselves individually and in small groups and then examined how a student approached the same problem through video or written children's work. More specifically, TCs first solved the problem for themselves and then solved it again anticipating how an elementary student might solve it. Solving the problem and discussing multiple solutions with their peers provided opportunities for TCs to develop awareness and appreciation of various strategies (correct and incorrect) that were often different from the ways in which TCs first solved the problem. Next the TCs analyzed a video clip or sample of children's work and were asked to document the children's thinking. This task required TCs to listen carefully to how children approach and solve a problem and to move beyond attending only to whether or not the answer is correct. Finally, TCs were asked how they, as teachers, might respond to the child. They were asked to consider the kinds of questions that could be asked of students, the kind of feedback given, and how the problem might inspire further mathematical extensions. Asking these questions, I hoped, would lead TCs to consider children's mathematical thinking as a resource for pedagogical decision-making. Similar to the use of manipulatives to support children's conceptualization of a mathematics problem, children's thinking was the "manipulative" used to support TCs' conceptualization of pedagogical practices.

A video excerpt from the research-based CGI collection serves as an example. This particular clip is approximately 2 minutes in length and highlights a primary level student's solution to the problem: "Naomi has three pockets. In each pocket there are six buttons. How many buttons does Naomi have?" TCs listen to the interviewer, pose the problem to the student, and then solve the problem themselves as they anticipate the young student's response. Many TCs consider this problem too difficult for primary level students and what they see in the video appears, at first, to confirm this assumption. For approximately 40 seconds after hearing the problem the child is silent, spends some time looking down, turns to the left, talks to herself and then shares the answer of "18". I often stop the video after 20 seconds to ask TCs what they think. Many say the student doesn't know the answer and should be told how to solve it, others conclude that the student is struggling with the question and suggest it is cruel for the teacher/interviewer to sit in silence without providing some support. All TCs are surprised when the student provides the correct answer and even more surprised with the communication of her nontraditional strategy. The video problem provides a context for challenging TCs' assumptions about what counts as evidence of thinking or not thinking and extends TCs' expectations of what young children can do. Most importantly the analysis of the video clip provides a beginning space for developing what Featherstone (2007) called a "questioning stance". TCs were encouraged to ask questions to understand the students' thinking, to test their assumptions and conjectures against evidence from the video clip, and to ground their interpretations and projected decision-making with evidence rather than opinion. In this way TCs were encouraged to notice (Mason, 2002) students' thinking differently. Video clips, such as those of CGI, provide opportunities for learning to notice and for developing what Sherin (2001) terms a "professional vision of classroom events".

Over the past year my colleagues and I, as members of the collaborative self-study group MTEC, have focused our attention on exploring the nature of algebraic thinking. As a group we found that patterning and algebraic thinking was a challenging topic for TCs to consider mathematically and pedagogically. Therefore a collective goal was to design good pedagogical problems that would engage TCs in questioning, listening and responding related to algebraic thinking. In this way we hoped to support TCs in developing a stance of inquiry about algebraic thinking.

We met as a group, shared ideas and research papers and became more interested in teaching and learning algebraic reasoning across the grades. At the same time a colleague introduced me to variation theory through Marton and Tsui's (2004) research. Variation theory is a theory of learning that pays attention to the different ways in which a concept, situation or phenomenon is experienced. It originated from Marton's (1981) earlier research based on a methodology called phenomenography designed for mapping the various ways in which people experience different phenomena of the world around them. Variation theory is based on the assumption that we experience situations or concepts in different ways depending on the individual and the perspective from which they are viewed. In learning a concept we attend to some aspects of the concept and not others, some will be in the foreground while others will be in the background (Marton & Booth,

1997). Changes in what we are attending to – that is, moving features of a concept between foreground and background – can lead to changes in experiences and interpretations.

Using a theory of variation to support TCs in learning to question, listen and respond to students seemed promising. Learning to attend to the features of the problem, listen to students' ideas and respond to their thinking required fluid movement of attention and awareness to the problem, students, and the context. Moreover, what might be considered to be in the foreground for teachers might be in the background for students, and vice versa. So I introduced variation theory to MTEC members and those who had the time were able to read various articles and chapters (e.g., Al-Murani, 2007; Davis & Dunnill, 2008; Marton & Booth, 1997; Marton & Tsui, 2004; Runesson, 2006, 2008).

Discernment, Variation and Invariation

A theory of variation posits that learning involves the development of a capability to discern or notice critical aspects of a phenomenon. To do this requires experiencing something else to compare to the phenomenon. For example, for TCs to evaluate a mathematical problem as rich would require experiencing a problem that is routine and procedural. To notice a geometric pattern would require experience with patterns that are not geometric. Marton and Booth (1997) argued that being aware of experiencing a concept in different ways is an important aspect of learning. This means that in order for aspects or features of a concept to be brought to the foreground of our awareness and therefore discerned, experiencing variation in those features is required. Marton and Booth (1997) and Watson and Mason (2006) argued that discernment is more likely if it is experienced against a background of relative invariance. For example, in order for students to become aware of the relationship between area and perimeter, we might keep the perimeter of a rectangle invariant and ask students to find the area of rectangles with that fixed perimeter. Similarly we might fix the area of a rectangle and ask students to explore the dimensions and perimeter of rectangles with that area. If there are limited opportunities to explore how critical features are variant or invariant as compared to other features of the concept, then there are fewer opportunities for discernment.

Runesson (2008) provides a useful example. In order to help students develop generalized algebraic expressions of a relationship between two variables (e.g., the chocolate bar costs 5 kronor more than a toffee), they need to be able to recognize the key aspects of what makes a generalized expression in terms of what varies and what is invariant. In this case $C - T = 5$ (where C represents the cost of the chocolate bar and T represents the cost of the toffee) can be expressed in multiple ways. However, the variable and equal sign are two concepts which students typically have difficulty understanding (Kieran, 1981; Knuth, Stephens, McNeil, & Alibali, 2006; Usiskin, 1988). In Runesson's (2008) example all the permutations for $C - T = 5$ were listed to indicate how the same example could be represented by different algebraic equations that include: $C-5=T$; $T+5=C$; and $C=T+5$. Comparing the lists of equations and systematically re-arranging the variables and equal sign focuses students' attention on the critical features of the algebraic expression. Noticing what changed and what remained invariant in these examples, and that an expression such as $C+T=5$ is not included, could lead students to a relational understanding of the equal sign and variable.

Variation theory therefore posits that experiencing a phenomenon in a new or different way can change students' awareness of its structure. Learning from the perspective of variation theory is therefore seen as "a process of differentiation rather than enrichment" (Runesson, 2008, p. 156). Learning in this way involves coming to know the differences and similarities between objects and identifying their properties. *Either* learning to discern or attend to what varies against a background that is fixed *or* learning to attend to what stays the same against a background that changes can increase the student's possible space of learning (Marton & Tsui, 2004).

My colleagues and I drew upon the principles of variation theory in our collaborative self-study to inform and develop pedagogical problems for teacher candidates. Marton and Booth (1997) referred to the phenomenon or something to be learned as the object of learning. Our intended object of learning was teaching and learning algebraic reasoning. This involved the recognition of patterns and the generalization of these patterns to algebraic expressions; how students learn algebraic reasoning; how these might be taught through open-ended problems; and how a teacher might pose problems, interpret student work and respond to students' algebraic thinking. It is possible to introduce variation theory to TCs as a pedagogical tool for designing learning opportunities; however that was not our focus at the time. Instead we drew upon variation theory as a way to help us, as mathematics teacher educators, learn more about TCs' conceptions of teaching and learning algebraic thinking. This chapter focuses on my experiences collaborating with colleagues to develop pedagogical problems with this goal in mind.

Data Collection Methods and Analysis

Data for this study are drawn from audio recordings and notes of the MTEC meetings, my own field and journal notes I had written during the MTEC meetings that focused on variation theory and algebraic reasoning, and the e-mails I had written during this time. I draw heavily on documentation collected on my own thinking. To deepen my own understanding of our collective work and my experience, I read through all of my notes, journal entries and e-mails looking for themes. With some themes conceptualized, I re-read the data with those themes in mind coding for further ideas, others I may have missed, and ones that did not seem to fit at all. As I analyzed what I had written I noticed that two themes were most prominent, one focused on the nature of collaborative self-study and one on the nature of variation theory.

Designing a Pedagogical Problem for Teacher Candidates

It was a challenge to design what we thought would count as a rich pedagogical problem for TCs. A mathematical problem appropriate for school students may not necessarily be pedagogically appropriate for TCs. Following Marton and Tsui (2004) and Watson and Sullivan (2008), who emphasize the importance of having a learning objective in mind while choosing the problem, we decided to design a task that would have the potential for prompting mathematical and pedagogical inquiry around teaching children algebraic reasoning. The task should engage TCs as genuine learners of mathematics and of pedagogy. It should inspire TCs to inquire and theorize about students' thinking, the potential of rich mathematical tasks, and the teacher's role. The task should also provide insight into TCs' thinking, inspiring us, as teacher educators, to theorize about TCs' reasoning and dispositions.

We spent a great deal of time experimenting with a problem called the Painted Cubes (If a 3×3 cube is dipped in paint, determine how many little cubes have paint on one side only; two sides; three sides; consider the same for a 4×4, 5×5 cube.) We video recorded school students in classrooms working on the problem. We also collected teachers' comments about their students' strategies, thinking and work on the problem. We compressed the digital recordings of classrooms into QuickTime movies so that all members of MTEC could analyze students' thinking and consider what might make a good pedagogical problem for TCs. Although the data collected were extensive, we found the Painted Cubes problem focused school students more on completing a required table than on articulating and communicating their reasoning and thinking. As a result there were few opportunities for TCs to examine the variation in student thinking within a classroom and across the grades.

We decided to try a different problem. We based this problem on a children's storybook and Chinese folktale called "Two of Everything" by Lily Toy Hong. The story is about a couple who discover a magic pot that produces double of whatever is placed within it. Two pennies go into the pot and four pennies come out. Three vases go in and six are produced. The storybook, we thought, would provide an interesting context for students to predict what could happen to different amounts placed in the pot, to develop a generalized expression, and to create their own algebraic expressions by developing their own magic pots.

To help TCs attend to the variation in student responses and conceptions across and within grade levels, we decided to read the same story and pose the same problem to students at three different levels: a Grade 1/2 class, a Grade 4/5 class, and a Grade 6 class. In addition, keeping the problem constant across grade levels provided TCs an opportunity to see how different teachers at various grade levels posed the problem to their students. At each grade level school students were involved in noticing and extending a pattern, as well as generalizing and representing it algebraically. The school students explored what it meant to double. Keeping the magic pot as a doubling pot, students explored what changed and what did not as items were placed in the pot. At each grade level, students also explored variation in developing algebraic expressions when they created their own function rules for their own magic pots. We video recorded the classrooms and gathered copies of student work. We compressed the video into QuickTime movies and made electronic copies of the students' work for MTEC members.

Our intention was to meet as a group and collaboratively analyze the classroom video records and school student work in order to design a pedagogical problem for our TCs. However, as the end of term approached and with little time for multiple meetings, I offered to analyze the collected video data and develop a possible problem for the MTEC group to consider. Watson and Sullivan (2008) suggest that choosing rich tasks to foster supportive mathematics classroom cultures requires consideration of these four elements: (1) level of student *choice* such as focus, approach and difficulty, all of which can contribute to motivation; (2) potential for prompting *communication*; (3) degree of *risk*, attending to variation in student responses to challenge, uncertainty and failure; and (4) level of potential *engagement* where the students view the task as purposeful and meaningful (p. 113). Applying these considerations to working with TCs meant that the problem needed to be one that could easily be posed in the elementary classrooms that TCs would eventually visit.

Findings

Mirroring Variation in Preparing Tasks for Students and TCs

Developing a pedagogical problem for TCs required identifying the critical features of the object of learning: teaching and learning algebraic reasoning. I expected the task to be more complex and challenging for some TCs than for others. Their own mathematical understanding of the problem and their experience of or disposition toward listening to students would frame what they attended to. The critical features of the task focused on the experiences of classroom students and on their teachers. Attending to school students involved foregrounding the variation in students' understanding of algebraic reasoning, particularly related to building function tables and developing generalized rules as algebraic representations. Attending to teachers' experiences involved foregrounding responses to student thinking, particularly how and why the teacher responded and the actions that followed. I began with the video recording of the Grade 6 class and watched it multiple times. I edited the recording down to a 10-min clip that included an excerpt of the teacher reading the story to the class, six Grade 6 students exploring the nature of the doubling pot, and a range of students creating their own magic pot rules. I selected clips of students explaining their own magic pot rule to provide evidence of variation in student thinking on the same mathematical problem. A transcript of the lesson was also developed.

I initially wanted to develop a task around interpreting and listening to students' thinking. However, as I analyzed the video and transcript I found myself drawn to the questions asked to students by the MTEC member filming in the classroom. She was an experienced teacher, very familiar with the story and the mathematics problem and so her questions posed to students were quite interesting. I communicated this thinking to MTEC participants:

> As a collection of clips for the Grade 6 class I see this as an opportunity to explore the kinds of questions a teacher might ask to learn more about what the students are thinking/understanding and the kinds of questions that could be asked to extend students' thinking. The problem seemed familiar to students (i.e. working with input and output functions was not new to them) and thus the problem was in some ways more an application of their thinking in a different context rather than a non-routine problem. However the questions the teacher asks are excellent and could be a good context to explore how we/TCs might respond to students to prompt/extend their mathematical thinking. (e-mail to MTEC members)

A goal was to develop a rich pedagogical problem that could be a place for TCs to explore both the mathematics of the problem as well as the pedagogical decision making involved. Thus, I reasoned that the video clip could also be a place to explore the practice of questioning or posing questions to students. The 10-minute video clip provided examples and variation of the kinds of questions that could be asked to students working on the problem. I developed a task for TCs that began by introducing the storybook and problem to them followed by analyzing the video clip of six students explaining the magic function pots. The focus of analysis of the video was on how a teacher might respond to students' comments, questions and ideas.

I developed the following "big ideas" of teaching algebraic reasoning related to this pedagogical problem and the selected video clip:

- Considering how a teacher might respond in order to (1) learn more about what students are thinking and understanding and (2) extend students' mathematical thinking;
- Exploring, adapting and extending a constructed problem for different grade levels;
- Predicting students' strategies for solving function table patterns and writing generalized expressions;
- Building students' algebraic thinking through designing questions that focus students on noticing patterns, extending patterns, writing generalizations, building T-tables, writing generalized algebraic equations; and
- Exploring the use of children's literature as a context for mathematical and pedagogical problems.

Then I selected the following excerpt and sent it to MTEC members. I suggested we could stop the video at the places I had annotated to ask TCs how they might respond to the student and why the teacher responded as she did. MTEC members responded by suggesting we provide an opportunity for TCs to experience the mathematical problem as their students would. It was suggested that TCs should play with the mathematics problem to build their own magic pots and exchange their pots (algebraic functions) with peers. Modeling the pedagogical problem in this way would certainly make it meaningful and practical. However, moving beyond this to include an examination of the pedagogical practices of questioning, listening and responding to students was also important. The excerpt below was selected as an example of a place where TCs could examine the kinds of questions they might ask students as they worked on the task. Although I include my analysis of the excerpt made at the time I developed the task, for TCs the analysis column was left blank to allow them to write their own analyses. My analysis includes the kinds of questions that could be posed to TCs.

	Transcript	Cynthia's analysis
S3	student writing table...	The student is completing an input output table with values (1, 5) (2, 8) (3, 11)... T points to the output numbers of 5, 8, and 11 to draw the student's attention to what is varying and how it is varying. Given this situation/context what question might TCs ask this student rather/other than this question posed.
T	*What do you notice about those numbers, here?... (pause) Do you notice a pattern at all?*	
S3	Yeah.	
T	What do you notice?	
S3	... (long pause) plus 2 each time.	
T	plus 2?	
S3	Oh... (long pause) this is plus 3, and this is plus 4.	S3 states from 5 → 8 is plus 3; from 8 → 11 is plus 4
T	So, 8 to 11 – is how many?	S3 corrects himself with teacher prompting
S3	8 to 11 is 3.	T drawing students' attention to the pattern in the output. Prompting S3 to predict the pattern. What question might TCs pose here? Could ask: Why do you think this teacher asked this question? If you were the teacher what question would you ask? What would you ask next, why?
T	*Okay. So, there's a difference of 3 between each of those. So, what do you think the next number should be... following that pattern...*	
S3	14.	T modeling how to check a possible answer.
T	14... So, does that work? 4 times 3 is 12, plus 2 is...	
S3	14.	Interesting question. T now trying to focus S3's attention on the relationship between input and output and the increasing pattern of the output. This can be very challenging for students. What is the T thinking here? Why this question? What is the T trying to accomplish with this question? Who is the question for – the T or the S? What question might you ask as the T to S3? What do you think about the last comment made by the T?
T	*14. Okay. Why do you think it's going up by "plus 3" over here?*	
S3	... (long pause) It was "times 3", so...	
T	Hmmm.... You keep thinking about that...	

Imagined Possibilities: What If?

This excerpt suggests multiple places where sameness and variation can be used as a tool to understand the pedagogical and mathematical decisions made. At the time the task was designed for TCs, these were not as apparent to me as they are now. With a focus on the teacher in the excerpt, a discussion could follow with TCs around possible ways a teacher can emphasize the dimensions of variation in learning to develop algebraic expressions from pattern rules. One of the challenges for students is to identify both the increasing pattern in the output and the relationship between the input and output values. In the case above, the teacher's question "Do you notice a pattern at all?" was an attempt to move the student's focus *from* creating output values based on the input value *to* creating output values based on the previous output, that is, to noticing that the output values were increasing by 3. The sameness focused on increasing by 3 or developing an algebraic expression that could represent "times 3 plus 2" by varying the input value. This led to an apparent contradiction for the student: *adding* 3 to the previous output value gave the next output value while *multiplying* the input value by 3 (and then adding 2) also gave the corresponding output value.

In a parallel fashion the excerpt could be used for teacher educators to consider how they might use this kind of task to emphasize dimensions of variation in the practices of listening to students, responding and posing questions. Watson and Sullivan (2008) described how 'What if' questions can prompt the dimensions of variation. In the task for TCs the questions that focused on the teacher's questions and dialogue with the student provided a context to explore variations in problem posing and response. What kinds of questions could be asked of a student in this situation, what else could be asked, and if we asked a different question – what response (or outcome) might be anticipated? These questions could help TCs attend to developing questions that not only ask students what they think, or to explain their thinking, but also emphasize the dimensions of variation. For example, a question asked by the teacher to many students was "Can you tell me about your magic pot". This question was the same and asked in the same respectful manner to many students yet the conversation that followed was different. Providing opportunities for TCs to see how an experienced teacher engaged students in discussions about their work could be key for TCs to imagine possible student responses and imagine possible next questions. Thus the task included some level of choice for TCs in that they could play with various imagined responses of the teacher.

Collaboration and Developing a Pedagogy of Inquiry as Teacher Educators

The discussions of the MTEC group around the development of a task/lesson for TCs that focused on the mathematical concept of algebraic reasoning were rich. MTEC members commented on how the discussions and meetings were unlike any other form of professional development they had experienced. Members welcomed the opportunities to work and learn from each other. In this case our collaborative self-study continued from developing the task to trying it out in my classroom with TCs. The opportunity to see and experience another teacher educator teaching TCs was very revealing. As one MTEC member stated "I'm involved in other forms of reflective practice in learning to be a teacher educator, but this experience of designing a task together and trying it out in classroom is so valuable. It has really helped me think more about a pedagogy of inquiry rather than teaching about teaching methods". For me the experience provided a context of exploring with my colleagues ways to imagine other possibilities for engaging TCs in mathematics teaching and learning.

Conclusion

This chapter examines the use of variation theory in the context of a collaborative self-study group, MTEC, developed for the purposes of improving teacher candidates' understanding of mathematics teaching and learning. Specifically, the chapter examines the pedagogical decisions made by my colleagues and me in developing and posing a pedagogical problem for use with teacher candidates. The potential of variation theory in the context of collaborative self-study was also examined and found to be useful for teacher educators in designing pedagogical problems and for promoting more complex ways of experiencing the teaching of algebraic thinking. However it could also be useful for teacher candidates in designing mathematical problems for their students. Thus, there are multiple levels in which variation theory, with its focus on sameness and difference, can be explored. One level involves TCs using variation theory to explore and examine mathematical concepts for the critical features that might be varied or remain the same. This could support TCs in their design of mathematical problems that will help school students deepen their understanding of a concept. For example, perhaps TCs could begin with mathematics textbook problems and adapt these problems so that a concept is introduced through attention to sameness and variation. Another level involves teacher educators using variation theory to examine pedagogical 'big ideas' for critical aspects that might be varied or remain the same. This, in turn, could support TCs to develop ways of posing good mathematical problems to their students, listening to student ideas and responding to their students' thinking.

The use of variation theory for teaching about teaching mathematics is becoming an important aspect of our MTEC meetings. We have found that many TCs have not yet developed a strategy for selecting mathematics problems to pose to their students. Nor have they developed ways of adapting or extending problems, or ways of using examples to help students attend to the critical features of a concept. Many TCs were surprised to learn that choosing good problems requires thinking deeply about the examples and problems teachers pose to students. My colleagues and I wonder about this and ask how we can use variation theory to support TCs in learning the relationships between the critical features of a mathematical concept so that they can pose good questions to their students. We also wonder how variation theory can continue to support us in the design of pedagogical problems for TCs. This chapter highlights potential in the use of variation theory. However, it also raises some questions.

First, it is a challenge to explore the multiple levels of variation theory in use. Considering how we might introduce variation theory to TCs in a mathematical context and at the same time use it ourselves as teacher educators within a pedagogical context is difficult. Supporting TCs to identify variation in possible conceptions of a phenomenon is challenging for those TCs who consider teaching as posing problems from the mathematics textbook and responding to students with answers from the teacher's guide. We will need to explore ways of making a problem or concept accessible to TCs, while they are trying to develop their understanding of teaching and learning. Further collections of video cases and examples of students working on problems and articulating their thinking will help TCs develop their understanding of students' variation in understanding of a concept.

Second, I wonder how useful variation theory is for teacher educators who find themselves teaching more *about* mathematical or pedagogical problem solving than *through* problem solving. Time constraints in an already very full course often lead to more teaching as telling than teaching as inquiry. Developing a pedagogical problem such as the "Two of Everything" problem is a great deal of work that involved time collecting video records of school students working on a problem, editing the video data to produce a video case, and designing a task for TCs with the collected artifacts. Although MTEC members found the process rewarding and stated it influenced their conceptions of teaching as teacher educators, it remains to be seen if the process can be sustained over longer periods of time or over multiple problems.

Third, developing and creating worthwhile problems and tasks for TCs is critical. With competing demands and extreme time constraints, mathematics teacher educators have little room for working with tasks that are limited, or too particular, in their learning opportunities. Variation theory provides a tool for working with colleagues to deepen our understanding of teaching and learning as mathematics teacher educators.

A next step for my colleagues and me will be to collaboratively design an assessment of TCs' pedagogical and mathematical understanding of algebraic reasoning. A significant aspect of using variation theory, as seen in work of Marton and Tsui (2004), involves researching students' initial and post-instruction conceptions of a phenomenon. For us this means researching TCs' incoming understandings of teaching and learning mathematics related to a particular concept, designing pedagogical problems based on these understandings, and then assessing TCs' understanding upon completion of the problem. MTEC will continue to be a context for exploring these uses of variation theory. It provides opportunities for us to observe each other teach the same concept to TCs and to observe TCs engage (or not) in the task. Although our current focus is on algebraic reasoning, it will be interesting to study if and how what we learn in this context will help us understand and prepare tasks for TCs for other concepts. Finally the process of collaborative self-study is a tool for professional development of mathematics teacher educators. This chapter provides one way of attempting to articulate my experience in a collaborative self-study as a mathematics teacher educator. It illustrates the possibilities of a collaborative self-study to understand and use a theory of variation to deepen and frame a practice as a teacher educator.

Putting on my gardening gloves and returning to my garden, I imagine how this one section near the wisteria tree will look next year. The wisteria grows rapidly, the irises bloom large but early, and I can hardly see the black mondo grass through the now-faded daisies. Bloom time, growth rate, maximum height, sun preference and soil conditions need to be considered. Taking these variables into account helps me decide what to keep and what to change next year. These provide growing possibilities. I know choosing plants for my garden will involve more than selecting those that look good at the gardening centre. Similarly, choosing or developing pedagogical problems for TCs involves more than selecting a mathematical problem that TCs can eventually use with students in their future classrooms. Although such criteria are important to consider, choosing a good pedagogical problem will need to involve teacher educators in exploring the critical features of a phenomenon that can be varied in relation to other features. This, I expect, could lead to more complex understandings and awareness of what it means to teach mathematics to teacher candidates.

I look forward to my continued work as a member of a mathematics teacher educator collaborative with the expectation that such work will help me learn more about the mathematical and pedagogical 'big ideas' for teaching TCs and the variation in ways TCs experience tasks related to these ideas. I am, in this way, developing a "professional vision of classroom events" (Sherin, 2001) as a mathematics teacher educator alongside my students and colleagues. These are, for me, growing possibilities at multiple levels.

References

Al-Murani, T. (2007). The deliberate use of variation to teach algebra: A realistic variation study. Doctoral thesis, Linacre College, University of Oxford, Oxford.

Bodone, F., Guðjónsdóttir, H., & Dalmau, M. (2004). Revisioning and recreating practice: Collaboration in self-study. In J. Loughran, M. Hamilton, V. LaBoskey, & T. Russell (Eds.), *International handbook of self-study of teaching and teacher education* (pp. 743–784). Dordrecht: Kluwer.

Carpenter, T., Fennema, E., Franke, M., Levi, L., & Empson, S. (1999). *Children's mathematics: Cognitively guided instruction*. Portsmouth, NH: Heinemann.

Carpenter, T., Franke, M., & Levi, L. (2003). *Thinking mathematically: Integrating arithmetic and algebra in elementary school*. Portsmouth, NH: Heinemann.

Coia, L., & Taylor, M. (2009). Co/autoethnography: Exploring our teaching selves collaboratively. In D. Tidwell, M. Heston, & L. Fitzgerald (Eds.), *Research methods for the self-study of practice* (pp. 3–16). Dordrecht: Springer.

Davis, P., & Dunnill, R. (2008) 'Learning study' as a model of collaborative practice in initial teacher education. *Journal of Education for Teaching*, *34*, 3–16.

Featherstone, H. (2007). Preparing teachers of elementary mathematics: Evangelism or education? In D. Carroll, H. Featherstone, J. Featherstone, S. Feiman-Nemser, & D. Roosevelt (Eds.), *Transforming teacher education: Reflections from the field* (pp. 69–92). Cambridge, MA: Harvard Educational Press.

Kieran, C. (1981) Concepts associated with the equality symbol. *Educational Studies in Mathematics*, *12*, 317–326.

Knuth, E. J., Stephens, A. C., McNeil, N. M., & Alibali, M. W. (2006) Does understanding the equal sign matter? Evidence from solving equations. *Journal for Research in Mathematics Education*, *37*, 297–312.

Loughran, J., & Northfield, J. (1996). *Opening the classroom door: Teacher, researcher, learner*. Bristol, PA: Falmer Press.

Marton, F., & Booth, S. (1997). *Learning and awareness*. Mahwah, NJ: Lawrence Erlbaum Associates.

Marton, F., & Tsui, A. B. M. (2004). *Classroom discourse and the space of learning*. Mahwah, NJ: Lawrence Erlbaum.

Marton, F. (1981). Phenomenography-Describing conceptions of the world around us. *Instructional Science, 10*, 177–200.

Mason, J. (2002). *Researching your own practice. The discipline of noticing*. New York: Routledge Falmer.

Nicol, C. (1999) Learning to teach mathematics: Questioning, Listening and Responding. *Educational Studies in Mathematics: An International Journal, 37*, 45–66.

Nicol, C., Novakowski, J., Ghaleb, F., & Beastro, S. (2010) Interweaving a pedagogy of care and pedagogy of inquiry: Tensions, dilemmas and possibilities. *Studying Teacher Education 6*, 235–244.

Roth McDuffie, A., Drake, C., & Herbel-Eisenmann, B. (2008). The elementary mathematics methods course: Three professors' experiences, foci, and challenges. In D. B. Jaworski & T. Wood (Eds.), *The mathematics teacher educator as a developing professional* (pp. 247–264). Rotterdam: Sense Publishers.

Roth, W. -M., & Tobin, K. (2005). *Teaching together, learning together*. New York: Peter Lang.

Runesson, U. (2006). What is it possible to learn? On variation as a necessary condition of learning. *Scandinavian Journal of Educational Research, 50*(4), 397–410.

Runesson, U. (2008). Learning to design for learning: The potential of learning study to enhance teachers' and students' learning. In P. Sullivan & T. Wood (Eds.), *Knowledge and beliefs in mathematics teaching and teaching development* (pp. 153–172). Rotterdam: Sense Publishers.

Sherin, M. G. (2001). Developing a professional vision of classroom events. In T. Wood, B. Scott Nelson, & J. Warfield (Eds.), *Beyond classical pedagogy: Teaching elementary school mathematics* (pp. 75–93). Mahwah, NJ: Erlbaum.

Stigler, J., & Hiebert, J. (1999). *The teaching gap: Best ideas from the world's teachers for improving education in the classroom*. New York: The Free Press.

Tidwell, D., Heston, M., & Fitzgerald, L. (2009). *Research methods for the self-study of practice*. Dordrecht: Springer.

Toy Hong, L. (1993). *Two of everything*. Park Ridge, IL: Albert Whitman & Company.

Usiskin, Z. (1988). Conceptions of school algebra and uses of variables. In A. F. Coxford (Ed.), *The Ideas of Algebra, K–12* (pp. 8–19). Reston, VA: The National Council of Teachers of Mathematics.

Watson, A., & Mason, J. (2006). Seeing an exercise as a single mathematical object: Using variation to structure sense-making. *Mathematical Thinking and Learning, 8*, 91–111.

Watson, A., & Sullivan, P. (2008). Teachers learning about tasks and lessons. In D. Tirosh & T. Wood (Eds.), *Tools and processes in mathematics teacher education* (pp. 109–134). Rotterdam: Sense Publishers.

Chapter 5
Resisting Complacency: My Teaching Through an Outsider's Eyes

Sandy Schuck

Studying My Practice

I have been a mathematics educator for all of my working life. My teaching has undergone some challenges, particularly at the time of my doctoral research. It was at that time that my beliefs about teaching and learning maths were disrupted and my practice changed substantially. I have documented this process in a number of articles (Schuck, 2009, 2006, 2002). The disruption led to some significant changes in my practice, which subsequently became embedded in my teaching. Now I have a new challenge to contend with: the challenge of complacency about my teaching. How can I creatively disrupt my current practices?

Self-study of teacher education practices is useful for resisting such complacency. Self-study aims at improving practice (LaBoskey, 2004) and reframing assumptions and practice (Loughran, 2002). This is often confronting as we struggle to step outside our own perspectives and comfort zones. One way of facilitating this process and of supporting our capacity to conduct self-study is through the use of critical friendships. Critical friendships tend to provide that very important mirror which shows us with all our flaws as well as our strengths (Handal, 1999).

Peer observation is often used synonymously with critical friendship. A description of peer observation follows that is aligned with my understanding of peer observation and critical friendships. Peer observation is a "collaborative, developmental activity in which professionals offer mutual support by observing each other teach; explaining and discussing what was observed; sharing ideas about teaching; gathering student feedback on teaching effectiveness; reflecting on understandings, feelings, actions and feedback and trying out new ideas" (Bell, 2005, p. 3). The features of this description that are critical for peer observation and critical friendship to work can be summed up by the statement that this is a collaborative activity conducted by peers for mutual benefit. What is not mentioned in this description, but is

S. Schuck (✉)
Centre for Research in Learning and Change, University of Technology Sydney, Sydney, Australia
e-mail: sandy.schuck@uts.edu.au

S. Schuck, P. Pereira (eds.), *What Counts in Teaching Mathematics*, Self Study of Teaching and Teacher Education Practices 11, DOI 10.1007/978-94-007-0461-9_5,

also important, is that both the observer and the observed learn about their teaching from the process.

Peer observation is often seen as having two possible purposes: development of early career teachers and performance management (Peel, 2005). In the case of this particular critical friendship, there is a third purpose for peer observation: the challenge to reconsider and reframe one's practice so as to maintain and enhance it. Where Bell and Mladenovic (2008) suggest a benefit of peer observation is the development of confidence in teaching, I suggest that in this case the benefit was the jolting of confidence and disruption of complacency. Ironically, peer observation seems to be mostly understood to be observation by a person with more experience or at a supervisory level, not about the mutual support of peers for each other's teaching. Critical friendship seems to be more applicable where peers are involved. For this reason, I tend to use the term "critical friendship" in preference to peer observation in this chapter.

Critical friendship has its own challenges. As Handal (1999) points out, the juxtaposition of the two words, "critical" and "friend" seemingly contains a contradiction. The first word indicates that the person is ready to find fault with our actions or teaching; the second that the person enjoys a close and supportive relationship with us and might be willing to disregard any deficiencies in our actions for the sake of our friendship. This discussion seems to indicate that critical friendship is complex and nuanced. What is important in critical friendship, however, is that someone whose opinion is respected by us is prepared to be honest about ways in which our teaching can be enhanced. Trust is essential in this relationship (Schuck & Segal, 2002) as is belief in the person's professional competence (Handal, 1999). So to be eligible for critical friendship, a person needs to be trustworthy and respected.

One aspect of eligibility for the role of critical friend that I test in this chapter is how well the critical friendship works when the friend is an expert in a different discipline area from the teacher being observed. A number of authors note the value in the critical friend being "an outsider" to the discipline or the culture in which they are operating (e.g. Handal, 1999). Handal notes that an "external critical friend" is able to see the situation afresh, without the assumptions that a person steeped in that discipline or culture might hold.

Handal's position is aligned with the one I hold. The reform movement in mathematics education, which was very influential in Australia in the 1990s, has powerfully influenced my teaching, as it has influenced that of many other maths educators. The emphasis on student activity, on moving away from "traditional" teaching of maths, and other aspects of maths teaching that arise out of the reform movement are likely to create strong differences between the teaching of maths and the teaching of other subjects. Accordingly, an external critical friend should be particularly valuable in encouraging maths educators to make the reasons for our "reform style" teaching explicit.

As well as seeing the value of having an "outsider" as a critical friend, I also see the value of having someone who shares my passion for teaching as essential for maintaining and enhancing my practice. I am extremely experienced in teaching mathematics education classes and have developed curricula for those classes that

I believe reflect good practice in teaching prospective primary school teachers how best to teach maths. I have had critical friends come into my classes over the last 10 years and have found this to be a beneficial process in supporting my reflections about my teaching. I, too, have attended classes of others, and I have learned as much about my teaching as they have learned about theirs during these peer observations (Schuck, Aubusson, & Buchanan, 2008). In fact, it is possible that as observer I learned more than when I was observed.

At the beginning of 2010, I had the opportunity to teach a maths education class again, after a break of a few years from teaching in the maths area. Mindful of the value of critical friendship in the past, I was eager to invite a critical friend to work with me in this subject. About this time, I had a discussion over coffee with a number of my colleagues in teacher education. These discussions are always fruitful, and we tend to solve the problems of the world, in particular educational problems. During this discussion an issue arose that is an enduring problem for those of us engaged in maths education. This issue concerns the unpopularity of maths in the eyes of a majority of students, at school and beyond (Boaler, 1997). I think I had made a comment about the elegance of a particular mathematical concept, and a colleague of mine, with whom I enjoy robust debate, suggested, possibly tongue-in-cheek, that the problem with maths curricula is that they are developed by people who are good at maths and love engaging in maths. His argument was that those not swept up by the beauty and elegance of maths will struggle to engage with the content we teach. For them, the elegance does not seem to be accessible, and a more "mundane" curriculum might have a better chance of succeeding with the majority of these students.

While this debate is beyond the scope of this chapter, it did provoke me to think that perhaps it would be useful to have a critical friend who was an "outsider" to maths education. I wanted to be able to have frank and full discussions about the approaches and content that I use, as seen by someone who is not a maths specialist. I anticipated that general discussion about pedagogy would probably be unaffected by our differences in discipline knowledge, but that I would gain valuable insights about my mathematical content and approaches as seen by someone not in the field. I wondered what unspoken assumptions I held about my students' knowledge and understandings, beliefs and motivation. I wondered what a teacher educator, with similar passions about teacher education but a different discipline expertise, would see that a colleague from maths education might not see.

A colleague in teacher education who teaches in areas of social and environmental education came to mind. This person (I shall call him Doug) would be more likely to experience my discussions about maths as an outsider, which might have some similarities to the way my students experienced the discussions; yet he would have the professional competence and understanding of pedagogy to enable useful critique. I have used my students as critical friends in the past, but I am mindful that they can only see part of the picture at the time of classes and that they might possibly have a different set of priorities to mine. For one thing, they will be participants in an assessment system which is likely to drive some perceptions and motivations to offer feedback. For another, some students might have a limited view of the value and power of maths. As well, they are at the start of their teaching careers and so are building their pedagogical know-how. I felt, therefore, that a colleague as critical friend would be a useful addition in my exploration of my teaching. Together with feedback from the students, which I gain in a number of ways through the semester, his comments would help me reframe practices with a view to enhancing them.

Coming to Grips with Maths Content

Maths education classes for prospective primary school teachers have a set of challenges that are probably unique to teacher education in this particular discipline. There are numerous studies which indicate that students enter teacher education programs with views of maths as "absolutist, rigid and taught by transmission" (e.g. Southwell, White, & Klein, 2004, p. 197). Added to this, a study of primary teachers in Australia showed that primary teachers indicate a lack of preparedness to teach science and mathematics (Angus, Olney, & Ainley, 2007).

It has also been shown (Boaler, 1997; Charalambous, Panoura, & Philippou, 2009) that learners of maths are often negative toward the subject. This negativity is often experienced by pre-service teachers as well (Foss & Kleinsasser, 1996; Pereira, 2005; Schuck, 1996; Schuck & Foley, 1999), particularly prospective primary school teachers. It becomes imperative, therefore, for teacher education programs in maths education to focus on: changing students' views on the nature of maths; increasing mathematical content knowledge for student teachers; and improving students' attitudes to maths. Charalambous et al. (2009) argue that inducing such changes in pre-service teachers' beliefs and attitudes is difficult yet feasible. It is this imperative that creates the set of challenges for maths education that teacher educators have to meet. I believe (but am ready to be corrected) that these issues tend to be more apparent in maths education than in other discipline areas in primary school teacher education. Consequently, these issues may have bearing on both student feedback and the observations of critical friends from other disciplines, a point I shall return to later.

It is with this knowledge and a desire to meet these challenges head-on, that I have developed the maths education subjects in our primary teacher education program in the Australian university in which I work. There are three maths education subjects; in recent years I have been responsible for the development of the first two in the sequence, and the third one follows a structure of content and approaches originally developed by me. After a break from teaching in maths education over the previous year or so, I returned to the maths education sequence in 2010. I taught the first subject in the sequence in our Autumn Semester 2010 (February–June), and I invited Doug to attend classes in that subject as a critical friend.

The first subject occurs in students' second year of the teacher education program. It introduces discussion about the nature of maths and about the changes that have occurred in maths education over the last decade. It models ways of learning maths collaboratively and I am careful to make explicit my reasons for running my classes in the ways that I do, with a strong emphasis on experiential hands-on activities, collaborative activities and discussion to justify and argue for mathematical results. I discuss, at some length, the learning theories that underpin my subject and drive my approaches.

The subject also covers problem solving as an approach, and alerts students to the way a problem-solving approach will be used in the class. The major mathematical content covered concerns measurement across a number of attributes such as length, area and mass; we focus both on the student teachers' learning of the concepts and also on developing approaches to teaching these concepts. The subject, therefore, teaches both about approaches to teaching and learning maths, and mathematical content matter. Another characteristic of these classes is that I like to find out about the mathematical backgrounds, attitudes toward maths learning and critical incidents experienced by students prior to entering the teacher education program. I believe that knowing their perceptions helps me to work with them on developing their competence and confidence in mathematics. My major aims for the subject are to support students' learning in a safe environment; to build their knowledge of mathematical concepts; and to help them develop some confidence in their understanding of maths to improve their attitudes to the subject.

The subject offering has two assignments and an examination at the end of the subject. In the offering in 2010, the first assignment was aimed at introducing the importance of maths in our daily life as well as showing how necessary it is to understand maths to get to grips with social issues. Students were required to find an authentic set of mathematical facts about a social issue. There were two examples that I provided for the students. One was the amount of water used in average households by different devices, a topic which is very important in our drought ravaged country, where much discussion is about how households can reduce water use. The second example concerned population growth, another topical issue at present with a debate raging about what number of Australians can be sustained by the country in the future. Students were required to pose a set of mathematical questions about the facts they found and provide worked solutions to their questions. The activities needed to be suitable for the last year of primary school, typically students who are 11–12 years old. They also needed to use a problem-solving approach, which aligned the assignment with the work the class was doing on problem solving.

The second assignment covered work in measurement by asking students, in pairs, to design and provide a workstation for area or time. The class would rotate through the workstations, both doing the activities at each station and also evaluating their peers' workstations according to a number of criteria. The assignment aimed to provide them with useful resources for their teaching, as well as developing their understanding of the pedagogical content knowledge needed to make the workstation appropriate.

The exam covered all aspects of the subject and required short answer or paragraph responses on how to teach different aspects of measurement, analyses of mathematics education today and calculations that would indicate content understanding. The exam does present some dilemmas for me: I see its worth as the sole assessment that is done under supervised conditions, but I am aware that it increases angst amongst the students as they prepare for it. I do believe it has pedagogical value however, as it encourages students to engage thoroughly with the subject matter.

The three assessment tasks are designed to: provide students with a sense of the value of maths in everyday life; give them confidence in preparing mathematical tasks and resources very different to those they might have encountered in their mathematical schooling; and give them an opportunity to develop their mathematical content knowledge and their pedagogical content knowledge.

The Critical Friendship

Doug is a social and environmental educator dedicated to teaching. He has invited me into his classes, and I have watched him teach students, both in core curriculum subjects and electives. In turn, in the past he attended one session of a maths class that I taught, and a few sessions of a research class I was teaching. Last year we team taught a final year elective subject providing students with skills and competencies to manage the day-to-day tasks of classroom teaching. However, it is noteworthy that while we were team teaching, we did not generally act as critical friends to each other, perhaps because we were so preoccupied with the teaching of the class that we did not want to spend time discussing each other's practice, but possibly because it was a shared venture and we were hesitant to comment on what the other person was doing. It appeared that critical friendship would be quite delicate in this context.

Doug and I have had many conversations about our teaching and often test ideas by discussing them with each other. Our critical friendship meets the criteria for critical friendship discussed earlier – we have a strong professional relationship, trust each other in terms of our professionalism and integrity and, I believe, we have a shared view that the other person is professionally competent in matters of pedagogy. As well we share a belief that doubt is to be valued in teaching, and I think we both see complacency as an enemy of quality teaching. However, we do have differences of opinion about aspects of teaching. Elsewhere we have used the metaphors of Doug as tour guide and me as travel agent (Schuck et al., 2008) – Doug takes an active role in showing the tourists the exciting sights whereas I like to set up the trip and facilitate the arrangements leaving the tourists to visit at their leisure. We enjoy debating the role of teacher as performer. While I can see that the students would enjoy having their teacher "perform" for them, by being dramatic, taking on roles, making jokes and generally being entertaining, I do not believe that this necessarily results in deep learning by the student. Doug argues that he wants his students to be thinking about his classes long after they are finished and these actions will help them to do so. I argue that students working on concepts, discussing them and developing strategies to problem solve will result in stronger learning, even if the students do not find it as appealing. We have not reached a resolution of this argument; nor are we ever likely to. I doubt that either of us has the correct answer.

In this semester, Doug had a light teaching load and so was happy to spend time attending my classes. The classes were three-hour long sessions, held weekly over 10 weeks. The 10 weeks were not continuous. In the middle of the semester students did not attend university but instead gained professional experience in schools. Doug attended five of the ten sessions so it was a sizeable amount of time that he dedicated to being a critical friend. I introduced Doug to the class in the first session and explained that we are critical friends and that he would be attending my sessions regularly.

When Doug came into class at each of the sessions he attended, he would seat himself at the back of the class and make notes. When the students worked on activities in groups, he would rotate amongst the groups, joining in their discussions and sometimes suggesting ways of solving the problems. After the class, we would discuss his thoughts and comments about the session.

Sometimes critical friends "set each other tasks": teachers ask their critical friends to observe an aspect of their teaching that they wish to enhance, or that they feel is problematic. We did not operate in this way. Our major intent with the critical friendship was to: share teaching ideas; develop awareness of how others might teach; stimulate thought about our own practices; and create opportunities for us to collaboratively reflect on, and discuss, what lies behind the teaching that occurs. Added to this, was my desire to get a sense of how a non-mathematics educator might perceive my practice and challenge my assumptions about maths teaching and learning.

Having a Critical Friend in the Maths Classes

In this section, I will discuss the main ideas that arose from Doug's observations of my classes and the ensuing discussions we had throughout the critical friendship. These ideas concern: my feelings of nervousness before classes; my efforts to build a safe and welcoming environment for the students; the relative importance of our different subject areas; the aims we each hold for our classes, together with our perceptions of our roles; and issues of control.

I have to confess that I am always slightly nervous before a class session, even though I have been a teacher educator for over 20 years. This nervousness was increased by knowing that Doug was coming into my classroom. Fortunately most of the time we had made arrangements for him to attend well in advance; but when he asked whether he should attend a session, just prior to the session, my first instinct was to say no. However, what I usually did was tell him that it was his decision. Doug conscientiously attended all sessions that it was possible to attend: half of the semester sessions.

It is useful to think about why I was nervous when he came into my classes. I have had critical friends in the class before, but sustained attendance raised the stakes for me. If he only came to one class, I could always delude myself, and perhaps him, that if the students were not engaged, or I was not happy with an aspect of the class, this was only one class and the "great" teaching had taken place in another class. Ironically though, I had invited Doug to be a critical friend in the first place, expressly to increase my nervousness and disrupt my complacency.

I think that part of my nervousness stemmed from my wanting Doug to affirm the way I taught. However, I feared that what I was doing, and the reasons for conducting the classes the way I did, would not appear to be all that impressive to an onlooker. I tend not to do anything showy but focus rather on student participation in the maths tasks to get the students learning. After all our conversations about teaching, I did not want Doug to think I was a fraud. Doug's presence certainly did disrupt my complacency and led me to think carefully about my rationale for conducting my classes in the way I did, according to my beliefs about the centrality of the students rather than the teacher in the activities.

The next issue that arose concerned my desire to provide a safe environment for students to learn maths. In the first session, I introduced students to the class by getting them to discuss their thoughts and experiences in maths education. About a quarter of the class offered comments; these comments were unanimously about the lack of confidence they felt or their confessions that they had always found maths hard. These types of responses are not unexpected; both my prior research and the literature indicate these are common reactions (Schuck, 1996). Because this negative response is a predictable outcome of my question about their past experiences, it provided me with an opportunity to create that safe classroom environment by assuring them that the work will be accessible to all and that they should send themselves positive messages by seeing the classroom as an encouraging and supportive place in which no question is regarded as stupid.

In our conversations after the class I was taken aback to find that my assertion – that I wanted defeatist comments about maths ability to have no place in our classes – was seen by Doug as being too forceful. He wondered whether my responses, when students said they were no good at maths, might be giving the impression that these statements were indeed "wrong answers". For example, my response to a student who said she never could do maths was that I anticipated that she would be able to do so in our classes and that she should not indulge such negative thoughts. On reflection, I understood the point Doug was making. Given that I meant my comments to be encouraging, this was useful feedback and made me more sensitive to the way my responses might sound to students who lack confidence. I needed to remember that feedback, no matter how positive, is interpreted in the light of what we already believe (especially about ourselves). So my assertion that the classroom was a safe environment might alert students to the fact that I anticipated problems, thereby confirming their beliefs that they were not adequate to the task. This discussion provided me with valuable insights that I might not have got from a critical friend who was a maths educator.

A major topic of discussion that arose out of Doug's observations centred on the relative importance of our subject areas. In particular, the dilemma of how much maths content to cover was highlighted in one discussion. We had just completed a session on collaborative learning, where the students rotated through a number of activities that were dependent on being collaborative. Doug said he wondered if we "were just doing maths for maths' sake". When I pushed to hear what he meant by that statement, he suggested that perhaps the students could do fewer activities and still understand the value of collaboration. I explained that I did have two goals for those activities, the one being to highlight the advantages of working collaboratively, the other being to engage the students in what I felt were interesting mathematical problems. Doug's comment concerned me because it seemed related to the earlier comment by the colleague about the enjoyment maths educators get from mathematical problems that might not be common to other groups.

Doug did however suggest that he might have felt that there were too many activities because as an observer he had less to do when the students were participating in those activities. In contrast, when I was holding whole class discussion, he was able to see and hear what was happening and was engaged in that. As well, as he was not actually engaged in the activities, he could not see that they were dealing with different mathematical concepts. So what looked like the repetition of an activity to him actually was developing different mathematical ideas for the students. Later, when reading a draft of this chapter, he added that he also feels a need to be proactive in class and "keep himself busy" when he teaches, which might have initiated this comment. This prompted me to think carefully about the students' engagement with the activities. I needed to satisfy myself that they were actively engaged in the activities and were learning new concepts. When I next take this class, I will think carefully about whether we can move on to different topics or whether the activities give rise to sufficient learning to justify the number of different tasks.

The other aspect of this discussion focused on the differing views we hold about the relative importance of our subjects. Doug feels that social issues permeate the lives of all of us, but seems to see the maths content that I teach as being situated in the classroom. I argue that to deal with those social issues, you need mathematical literacy; and I hope that the first assignment that I gave students provided evidence of this.

After discussion, we both came to realise that we have overlapping but different aims for our classes and hence for our teaching. My aim is to improve students' attitudes to maths, provide them with an opportunity to work with accessible yet engaging concepts and prepare them to be confident and competent maths teachers in primary schools. Doug's aim is to challenge them to be critical thinkers and good citizens. The preparation for teaching, while part of his goals, is not his first priority.

Doug argued that it would not be unreasonable to expect any functioning person in our society to be able to engage in the social issues he teaches them, but that it would be less common for our students to be mathematicians. He feels that he is giving students the tools to become active and effective members of society, while I believe that the mathematical processes I emphasise contribute to our ways of functioning well in society. These mathematical processes comprise questioning, developing strategies, justifying processes and solutions, communicating and reflecting and are part of the Working Mathematically section of the mathematics curriculum. I believe these processes are needed to become effective and competent members of society. These processes are strikingly similar to the ones Doug wants to develop; hence the overlap in our views.

I accept that my students might not be exposed to higher order maths concepts in my classes, but they are going to develop as critical thinkers by working mathematically as indicated above. So perhaps we are moving toward the same goal, in different ways. I do believe that if the students are confident and competent teachers of primary level maths, they will be more likely to venture further mathematically in their lives as teachers than they will if they dislike maths, fear it and wish to avoid it.

The activities also led to an interesting discussion about our roles in the classroom. Doug fears that he is not earning his keep if he is not carefully managing the activities, and playing a central role in their execution. I feel that my most successful sessions are those where the students are actively carrying out activities and I have little or nothing to do during that time, except be available if students want to ask me a question. Of course, this is not to say that there is no work involved in the classes – but much of that work occurs before the session, in my preparation so that the activities can run seamlessly. As well, I spend some time briefing the students on the activities and debriefing and sharing their understandings after the activities. The tour guide and travel agent metaphors spring to mind again. Doug teaches for impact: he wants students to go home and discuss the events of the class, and they do. I want my students to be impressed, not with what I have done, but with what they can do. I hope that I am succeeding in this.

I think that our different disciplines contribute to our thinking about control in the classroom. We know teachers' orientations to, and beliefs about, content influence the way that the content is taught (Ball, Thames, & Phelps, 2008). As mathematics educators we have had much exposure to reform ideas about teaching that indicate that students learn more maths when they, rather than the teacher, are actively in control. This principle guides the way that I conduct classes. I believe that educators in other disciplines have not been exposed to the same sort of reform ideas that maths educators have been, and that this influences our differing approaches to teaching. I do not know if Doug's students learn more from being engaged and entertained by his teaching than my students learn from struggling with ideas and learning from each other. Perhaps entertainment, engagement and struggling with ideas are all essential for learning. I do know from researching the experiences of graduates from our program that they "do" maths teaching differently from their predecessors, so change is occurring.

This discussion with Doug is ongoing; it may never be resolved. But even where I feel strong disagreement about comments he has made, such as the one above about the importance of social and environmental education as compared to maths, it gives me pause to think about the discussion, and to reflect on what that might mean for my practice.

Learning from Critical Friends

Toward the end of the semester, Doug indicated to me that "he was finally getting my teaching". I was both excited and dismayed. What had been happening prior to this? Had he been a polite observer who did not see much sense in my approaches despite arguing to the contrary? When I asked more about this, he told me that he was gaining a better understanding of the level of thought and planning that I gave to my classes, and the goals that I had for them. I believe that the differences in our teaching styles referred to above had initially stood in the way of that understanding. Doug suggested that his belief in the importance of the role of the teacher in stimulating thought may have prevented him from seeing that the students were engaged and were learning, although I was not in a central role. He added that he was seeing depths that he had not previously noticed. The discussion we had about this was valuable to both of us, and it was aligned to the desire for our critical friendship to provoke our thinking about what lies beyond the teaching rather than just examine the acts of teaching themselves. It highlighted to me, that I perhaps need to do more to be explicit about why I teach the way I do. If an experienced teacher educator was taking time to understand my teaching, how much more difficult would it be for my students?

Throughout the semester in which this critical friendship took place, I asked myself what our location in different discipline areas brought to our critical friendship. Initially, we both suspected that the differences in our teaching came, not from our different discipline bases, but from our different teaching styles. I am now beginning to see the impact that the reform movement in maths has had on my thinking and consequently on my approaches to teaching maths. This suggests that Doug and I are teaching differently because we are in different disciplines. Doug has asked me if I would teach his subject differently to the way I teach maths. I am not sure of the answer. He has suggested that he would be likely to teach maths differently from social and environmental education. We will never know the answers to this question, but it provides food for thought about the peculiarities of each of our subjects.

Doug's views and mine do differ about the importance of what each of us does in our classes. Doug believes that the students need to grapple with big ideas and that they should be able to do so. I argue that the maths I teach will assist our students to access these ideas, but my first priority is to support them in becoming competent and confident primary school teachers of maths.

I found that when we had discussions about the more conceptual aspects of our critical friendship, much of what Doug said indicated that he was looking at his own practice in a way that was stimulated by his observation of mine. Likewise, I found that his comments to me made me pause and think about my reasons for my approaches and content. I had to justify them, if not to him, certainly to myself, in the light of his comments. We agreed that we might not be questioning our own teaching if we did not have these opportunities to challenge each other. To improve our teaching we need to go beyond the comfortable.

I notice that one of the paradoxes in my critical friendships is that I engage in these friendships to unsettle and challenge me and to disrupt my complacency. And yet, what really drives me is a desire for affirmation. The affirmation is food for my soul, but the critique is the medicine to help me get better. I suspect most people in critical friendships have similarly ambivalent feelings about the process. To be useful, the critical friendship needs to be edgy and uncomfortable so that reframing of practice and assumptions can occur; but to be sustainable and manageable it also needs to affirm the participants. It is this notion of having both the "critical" and the "friend" that make the process so powerful.

So was my complacency jolted by this critical friendship? Indeed, it was. The critical friendship provided me with a wonderful opportunity to reflect, discuss and grapple with ideas about teaching and learning. I have learned to be more sensitive to my students' reactions to what I say to them, remembering that they have a lifetime of experiences to overcome. I have had an opportunity to revisit the nature of my role in the classroom and have had to justify what I do to both myself and to Doug. And I am encouraged to continue to stimulate my students' thinking about maths and the nature of the content and processes in which we engage.

I suspect too, that observing Doug's classes in a sustained way will also provide me with much food for thought. I noticed that Doug spent much of the time reflecting on his practice, stimulated by our discussions and his observations. I look forward to participating in his classes as a critical friend to gain added insights into my practice.

As Doug so eloquently put it, we might not have explored new territory but we have dug deeper in the existing territory. We have devoted much time and thought to our teaching and this probably would not have occurred without the critical friendship.

References

Angus, M., Olney, H., & Ainley, J. (2007). *In the balance: The future of Australia's primary schools*. Kaleen ACT: Australian Primary Principals Association.

Ball, D., Thames, M., & Phelps, G. (2008). Content knowledge for teaching: What makes it special? *Journal of Teacher Education*, *59*, 389–407.

Bell, M. (2005). *Peer observation partnerships in higher education*. NSW, Australia: Higher Education Research and Development Society of Australasia Inc.

Bell, A., & Mladenovic, R. (2008). The benefits of peer observation of teaching for tutor development. *Higher Education*, *55*, 735–752.

Boaler, J. (1997). When even the winners are losers: Evaluating the experiences of "top set" students. *Journal of Curriculum Studies*, *29*(2), 168–182.

Charalambous, C.Y., Panaoura, A., & Philippou, G. (2009). Using the history of mathematics to induce changes in preservice teachers' beliefs and attitudes: Insights from evaluating a teacher education program. *Educational Studies in Mathematics*, *71*, 161–180.

Foss, D., & Kleinsasser, R. (1996). Pre-service elementary teachers' views of pedagogical and mathematical content knowledge. *Teaching and Teacher Education*, *12*(4), 429–442.

Handal, G. (1999). Consultation using critical friends. *New Directions for Teaching and Learning*, *79*, 59–70.

LaBoskey, V. K. (2004). The methodology of self-study and its theoretical underpinnings. In J. J. Loughran, M. L. Hamilton, V. K. LaBoskey, & T. Russell (Eds.), *International handbook of self-study of teaching and teacher education practices* (pp. 814–817). Dordrecht, The Netherlands: Kluwer.

Loughran, J.J. (2002). Effective reflective practice: In search of meaning in learning about teaching. *Journal of Teacher Education*, *53*(1), 33–43.

Peel, D. (2005). Peer observation as a transformatory tool? *Teaching in Higher Education*, *10*(4), 489–504.

Pereira, P. (2005). Becoming a teacher of mathematics. *Studying Teacher Education*, *1*(1), 69–83.

Schuck, S. (1996). Chains in primary teacher mathematics education courses: An analysis of powerful constraints. *Mathematics Education Research Journal*, *8*(2), 119–136.

Schuck, S. (2002). Using self-study to challenge my teaching practice in mathematics education. *Reflective Practice, 3*(3), 327–337.

Schuck, S. (2006). Evaluating and enhancing my teaching: What counts as evidence? In P. Aubusson & S. Schuck (Eds.), *Teacher learning and development: The mirror maze* (pp. 209–220). Dordrecht, The Netherlands: Springer.

Schuck, S. (2009). How did we do? Beginning teachers teaching mathematics in elementary schools. *Studying Teacher Education*, *5*(2), 113–123.

Schuck, S., Aubusson, P., & Buchanan, J. (2008). Enhancing teacher education practice through professional learning conversations. *European Journal of Teacher Education, 31*(2), 215 – 227.

Schuck, S., & Foley, G. (1999). Viewing mathematics in new ways: Can electronic learning communities assist? *Mathematics Teacher Education and Development, 1*, 22–37.

Schuck, S., & Segal, G. (2002). Learning about our teaching from our graduates, learning about learning with critical friends. In J. Loughran & T. Russell (Eds.), *Improving teacher education practices through self-study* (pp. 88–101). London: RoutledgeFalmer.

Southwell, B., White, A., & Klein, M. (2004). Learning to teach mathematics. In B. Perry, G. Anthony, & C. Diezmann (Eds.), *Research in mathematics education in Australasia 2000–2003* (pp. 197–217). Flaxton, QLD: Mathematics Education Research Group of Australasia.

Part II
Examining Our Practice: Conflicts, Dilemmas and Incongruities

Chapter 6
How Students Teach You to Learn: Using Roundtable Reflective Inquiry to Enhance a Mathematics Teacher Educator's Teaching and Learning

Robyn Brandenburg

Introduction

As with many mathematics teacher educators, I purposefully set about challenging the curriculum and pedagogical status-quo; I want my Bachelor of Education pre-service teachers (PSTs) to experience learning and teaching mathematics in a connected and powerful way. My research of my teaching of the Learning and Teaching Mathematics courses at my university over the past 10 years suggests that the majority of PSTs initially fear teaching mathematics and need structured support in understanding the content of the primary (elementary) and secondary mathematics curriculum; they require ongoing exposure and related experiences linked to developing an understanding of mathematics pedagogy and they need guidance to understand and value reflective practice as a method to unpack their learning. An integral belief that underpins my new approach to my teaching in the Learning and Teaching Mathematics courses concerns the value of focusing on PST experience as a springboard for learning. Consequently, I initiated Roundtable Reflective Inquiry (RRI) sessions as a key practice that enables PSTs to reflect on experience in an ongoing, structured manner. These sessions (described in detail below) are primarily based on PSTs generating discussions based on critical events in their mathematics teaching practice and as such are driven by PST experience; my role as teacher educator is to facilitate, synthesise and challenge (often) firmly held assumptions.

The research question underpinning my self-study of my teaching practice – *how can PSTs learn about teaching through their own experiences rather than mine?* – emerged from my experience of teaching mathematics and was a natural extension of my pedagogical beliefs. However, as this chapter will reveal, an often unexpected outcome of pre-service teacher education is teacher educator enrichment, knowledge and transformation. As Kosnik (2001) suggests, "So often we, in teacher education, see ourselves as agents for our student teachers: motivating them, informing them, guiding them, preparing them. We do not think of it as a process that will

R. Brandenburg (✉)
School of Education, University of Ballarat, Ballarat, VIC, Australia
e-mail: r.brandenburg@ballarat.edu.au

S. Schuck, P. Pereira (eds.), *What Counts in Teaching Mathematics*, Self Study of Teaching and Teacher Education Practices 11, DOI 10.1007/978-94-007-0461-9_6,

also change and enrich us" (p. 65). While my intention, and indeed the focus of my self-study was to enhance PST learning and teaching in mathematics, it soon became evident that I was learning as a teacher educator and my new approach was contributing to my own pedagogical development as a teacher educator. While PST responses and engagement with RRI sessions provided opportunities for PST learning, my participation enabled me to see that which I had not previously noticed and in this sense, the PSTs taught me more about learning as a mathematics educator.

Learning and Teaching Mathematics I and II are two compulsory mathematics courses undertaken by first and third year Bachelor of Education students as part of their degree. Each of these courses consists of a lecture and a 2-hour tutorial once a week for one semester (12 weeks duration), and PSTs plan and teach mathematics lessons with a partner in local schools for 6 weekly sessions. On their return from their teaching experience, PSTs participate in RRI sessions which become the dedicated space to formally reflect on experience in a structured way. In this chapter, I explore RRI and through my systematic collection and analysis of data from RRI sessions show how I have come to understand more about the ways that structured and ongoing reflection can impact pre-service teacher and teacher educator knowledge and learning. I also came to understand more about the role that PSTs played in shaping my learning.

Why Roundtable Reflective Inquiry (RRI)?

How does a mathematics teacher educator involve PSTs in reflective practice that is meaningful and engaging? PSTs in my Learning and Teaching Mathematics Courses (2000–2002) suggested that reflective practice at best was a task that needed to be completed for the lecturer and to the lecturer's satisfaction. At worst, reflective practice was seen as meaningless and futile – as one PST stated, *What is the point of writing on the right hand side of the page what you wrote about on the left hand side... I usually write what I think the lecturer wants to read* (May, 2002). Yet another suggested that a focus on reflective practice was just *pyramids and crystals... it's just a fad* (August, 2003). My initial challenge was to not only identify and understand PSTs' assumptions about reflective practice but to encourage reflective practices that engaged PSTs and enhanced their abilities to critically reflect on their teaching and learning. A seemingly natural place to begin was to provide opportunities in my classes for PSTs to learn about teaching through reflecting on and unpacking their own experiences, rather than mine. At the beginning of my research, I restructured traditional tutorial sessions to include Roundtable Reflective Inquiry sessions (Brandenburg, 2004a, 2004b, 2008). My rationale for creating roundtable reflection sessions was twofold: (1) they were created as a response to research that indicated high levels of PST dissatisfaction and frustration with reflection as a meaningful practice in their Bachelor of Education program, and (2) they provided a structured space for unpacking experience connected with PST partner teaching in schools. Typically, PST partners plan and teach weekly mathematics lessons in schools and return to university to participate in roundtable sessions which are conducted for approximately an hour. PSTs separate into two groups (own choice) and arrange the tables in the teacher education classroom to reflect a "roundtable" structure. To initiate the RRI process, PSTs identify a critical incident, interaction or event that occurred in their teaching in schools and write about this for two minutes. They then briefly share the critical incident, interaction or event with the group (verbal sharing is crucial) and as a group, they prioritise which incident, interaction or event will be initially discussed in detail. One member of the group is then nominated to take (brief) notes of the discussion (an important data source). The RRI then proceeds with a detailed elaboration of the issue which might include questions related to the teaching context; why was this was an issue or problem; what was your reaction; and how did this make you feel? The group focus then is to explore multiple perspectives related to the issue or action and it is important that each PST has an opportunity to contribute to the discussion. For example, PSTs might ask if the initial action can be viewed in another way. What might be the advantages/disadvantages of alternatives offered from the group? If the situation were to occur in subsequent teaching, what approach might you take? PSTs came to establish patterns of questions to enable them to understand their learning more fully.

It is also important to note that these roundtable sessions are not formally assessed as part of the Learning and Teaching Mathematics courses. Roundtables were designed to capture learning based on experience; to enable PSTs to identify issues, engage in discourse and unpack teaching and learning based on their own experiences as mathematics teachers in schools. RRI was consistent with my seven teacher educator beliefs that:

1. PSTs and the teacher educator need opportunities to make sense of experience/s in a supportive environment;
2. PSTs generate discussion by raising issues related to their experience;
3. the role of the teacher educator is to introduce the session, clarify the framework and consciously refrain from dominating discussion;
4. PSTs should have opportunities to raise issues and hence develop their own voices;
5. learning outcomes cannot be predetermined;
6. learning/s are made explicit; and
7. opinions are respected.

Thus, roundtables were designed to allow for a focus on discussion as a reflective technique to elicit learning. Roundtable reflection engaged PSTs by encouraging them to problematize learning through identifying a critical incident, event or interaction and systematically examining this experience with the assistance of their peers – they raised issues related to *their experiences* and their needs as learners. However, as previously stated in the introduction to this chapter, much of what we experience and assume as mathematics teacher educators remains implicit, or tacit, and it was through designing, implementing and systematically reflecting on RRI sessions that I came to challenge my assumptions about reflection in general, and

RRI in particular. Although the roundtable sessions were specifically designed to encourage the development and articulation of PSTs' voices, they became a major catalyst for me in learning more about the less explicit aspects of my pedagogy (Kosnik, 2001).

Self-study Methodology

This research was conducted as a self-study (Berry, 2007; Brandenburg, 2008; Hamilton & Pinnegar, 1998; LaBoskey, 2004; Loughran, 2004, 2006; Loughran & Russell, 2002; Schuck, 2002). Self-study methodology has continued to gain significance as a powerful approach to educational research which has contributed collective insights into the complexities associated with teaching and learning about teaching in pre-service teacher education (Pinnegar & Hamilton, 2009). Self-study demands a close scrutiny of one's pedagogy and enables responses to dilemmas experienced within one's practice to be identified and enacted in opportune ways. As Pinnegar and Hamilton (1998) suggest, self-study is "not a collection of particular methods but instead a methodology for studying professional practice settings" (p. 33). Five distinguishing characteristics of research (LaBoskey, 2004) that guided my self-study of teaching practice were the following: (1) an initiation by and focus on *self*, (2) it was a study that was improvement-aimed; (3) it was interactive at one or more stages throughout the process; (4) I utilized multiple, mainly qualitative methods of data collection, analysis and representation and; (5) validity was achieved through the testing, sharing and re-testing of exemplars of teaching practice (in my case, RRI). I wanted to focus on the pre-service teacher *self* and in doing so, I came to understand more about my teacher educator *self*; I wanted to improve pre-service teacher mathematical knowledge and understanding and in doing so, I came to understand more about my own pedagogy and RRI provided the structured, interactive and reflective space needed for this new learning.

Participants

The research reported in this chapter is based on data drawn from a 3-year longitudinal self-study (2002–2004) conducted with cohorts of PSTs at varying stages of their Bachelor of Education Degree at the University of Ballarat, Victoria. The context for this self-study research was Learning and Teaching Mathematics I (First year) and Learning and Teaching Mathematics II (Third year). The average cohort size was 85 and each pre-service teacher cohort constituted both males and females – averaging 20–25% males; 75–80% females – the majority of whom were Caucasian. The percentage of mature age PSTs (i.e., those who are non-direct school leavers and have generally had workforce experience) represented approximately 20% of the total number of PSTs in each cohort. All PSTs referred to in this chapter have been allocated pseudonyms. My role was mathematics teacher educator participant/researcher.

Data Collection

A total of 17 roundtable reflection sessions were conducted as part of the self-study research over a period of 3 years. I employed a variety of methods for gathering data from these RRI sessions including: audio-taping and transcription; field notes (teacher educator); PST post-roundtable summaries and written reflections. Not all RRI sessions were audio-taped as I felt that this approach to data collection may have been invasive and confronting for some PSTs. If so, they could contribute to silencing individuals and therefore ultimately influence the interaction between PSTs during sessions. However, in my role as participant researcher, the audio-taping and transcription of the session/s allowed for an increased focus on my participation within the session and provided rich data which could be re-visited and analysed from multiple perspectives.

The transcription of the RRI data included a four stage process. Following the audio-taping of a roundtable session, the initial stage was the transcribing and organisation of the data; the second stage involved a form of tabulation whereby the transcript was then allocated a reference and a line number; the stage three involved coding whereby the transcript was coded and categorized; and the stage four involved the categorical analysis where each transcript was read to identify "relationships between data items" (Lankshear & Knobel, 2004, p. 271), which then were refined to a number of categories.

Challenging My Assumptions

My assumptions about the possibilities of RRI being a powerful approach to unpack PST teaching and learning emerged from my assertions (above). Two of my teacher educator assumptions related to RRI for the mathematics courses that I taught were as follows:

- Assumption One: Roundtable Reflective Inquiry will provide opportunities for PSTs to challenge taken-for-granted assumptions about learning and teaching mathematics, and
- Assumption Two: PSTs maximise learning opportunities by participating verbally in roundtable reflection discourse.

In what follows, I will examine these two assumptions about RRI, examine PST and teacher educator learning through the RRI, and discuss the ways that this new learning has impacted my practice. I start by discussing my first assumption linked to RRI and uncover how my new learning challenged this assumption and so altered my practice.

Assumption One: Roundtable Reflective Inquiry provides opportunities for PSTs to challenge taken-for-granted assumptions about learning and teaching mathematics

Jess – Teaching Subtraction – An Example of RRI Unpacking of Experience

During Roundtable 1, Jess described her problem as she experienced it during her teaching. She was attempting to teach about the concept of subtraction to a group of grade one students. Jess identified that her planned strategy was ineffective for the majority of her group:

> I had six in my group and one or two of them knew what the take-away sign [was] ... I assumed that they [all knew it] I used the felt tree I made in the teaching aids last semester, and I asked "Can anyone tell me what subtraction or take away means?" And they're all ..."it's like when someone steals apples" or they just had no idea ... and I thought ... they were just making up things [pre-service teacher laughter] it was things you have in a tree. They just had no idea ... I thought they would have a much bigger understanding of what subtraction was. (Jess, Roundtable 1, July 2002)

Jess had planned the subtraction lesson with her teaching buddy and her mentor teacher. While she was sharing her dilemma during the RRI session, her voice was faltering and she was clearly upset. Jess continued:

> I took the tree away.... they didn't have an answer of what subtraction was... I gave one sort of answer and then I sort of had... to explain... give them, yeah, tell them what it was and then a lot still couldn't [understand] so I put three apples [up] and said I was hungry and ate two and took two away and they couldn't even write... in their books, like three take away one equals two... They just, they just, I just thought they'd be able to do that pretty easy. I thought they'd be able to participate. (Jess, Roundtable 1, July, 2002)

In her articulation of this problem to the group, Jess displayed signs of frustration and agitation, which were evident in the quiver in her voice, the speed of her speech and her inability to complete sentences. Her frustration also revolved around the careful way in which she had approached her teaching; for example, she had met the classroom teacher prior to teaching the class to discuss the students' knowledge of the content area; she had volunteered time to meet the students in advance and she had planned the session and created teaching aids to assist in the teaching of the subtraction concept. In short, she had tried to do everything she could to make the lesson a successful learning experience. At this point during the session, I began to consider Jess' pedagogical content knowledge. She understood subtraction as "take-away", not realizing that there are further complexities. Did she understand that the symbol "-" could also be used in comparison or missing addend situations? Did she understand that these 6-year-old students may not have had the language to engage fully with the mathematical concept? Her student response linking take-away to "stealing", although humorous for the PSTs in the session should have alerted her to a mathematical language barrier. Where she thought that the students in her class should have had a much deeper understanding of subtraction, perhaps it was her own lack of conceptual understanding and indeed experience that confronted her during her teaching session.

The subsequent conversation with her peers during the RRI session provided evidence to show that Jess needed to consider alternatives and a number of alternatives were provided by the PSTs in the group. One of the alternatives included Sally's suggestion of the use of a counting book for student clarification, a modification of the terminology of "take-away" in order to assist student understanding through a visual stimulus and a description of her experience where the students created their own subtraction algorithms:

> On my last teaching round I had preps [6-year-old students] and I did subtraction and used my book that we made. We didn't work with subtraction, we used take away but my teacher suggested to allow them invent their own [subtraction algorithms]. We used apples and when they cut it, it was like you know when you invent spelling and they were inventing their own sort of little algorithms. A lot of them did things back to front and . . . some were just [using] a slash through the three and one take away three oh, three take any one or something, they'd do a three with a slash then a two and it was when you went to them they explained to you that you could see that they were totally on track. (Sally, Roundtable 1, July 2002)

Sally was reporting on a practice she had used during her placement based on an approach to teaching that encouraged students to develop and explore their own algorithms, under guidance, rather than be taught using specific terminology and structured steps. Sally's teaching of the concept of subtraction was in contrast to Jess's expectations for her teaching of the subtraction concept that included a linear process of developing a concept with a group of students and creating understanding by staging group conversations in a sequential manner. Sally, however, suggested a somewhat less widely acknowledged approach to teaching the concept of subtraction; i.e., by allowing students to explore the creation of algorithms using trial and error. While I was listening to Sally's experience and explanation, I began to consider that she may have experienced teaching and learning mathematics in a more powerful way than Jess. Jess may have left the RRI session questioning her own content knowledge and the need to provide multiple strategies when teaching mathematical concepts; Sally may have left the session buoyed by her contribution and I left the session knowing that the PSTs needed opportunities to unpack their learning in a structured, supportive environment. PSTs engaged in the RRI discourse; they recognised Jess' frustration and anxiety and they presented a range of alternatives for future action. PSTs were clearly unpacking their experiences and their assumptions were being challenged.

RRI has been embraced and endorsed by PSTs as an effective and engaging way of both problematising and unpacking mathematical learning based on experience. Analysis of all data sources across all cohorts suggested that RRI sessions were identified by PSTs as one of the most engaging aspects of the Learning and Teaching Mathematics units. As Sam stated, *[RRI] is good because we can really play to it, because they're our own problems they're not... a case scenario; we can find out different ways to sort out our problems in relation to schools and work... we don't do it anywhere else... like how to deal with problems* (Roundtable, 2003). During

RRI, PSTs challenge others' assumptions, and in doing so, identify and justify and/or reframe their own – *I learnt in this [Roundtable] format... I linked the things that others suggested and talked about my own opinions and ideas, reasoning... as to whether I agreed with what was said, or did not. I learnt about others' opinions and allowed these to influence my own* (Post-session written reflection, 2006) and they consistently unpacked pertinent issues – as Jay reported, *RRI enabled me to start getting down to the underlying issues concerning students' learning and associated needs [and] learning priorities* (Roundtable, 2005). In doing so, PSTs collaboratively challenged and interrogated their own and their peers' (often) taken-for-granted assumptions (Brookfield, 1995) about learning and teaching.

Teacher Educator Learning

My assumption that RRI would provide opportunities for PSTs to challenge taken-for-granted assumptions about learning and teaching mathematics was supported. However, my learning from this specific interaction during this roundtable became the springboard for much of my subsequent learning as a mathematics teacher educator. Following this session, I sent the audio-tape to be transcribed and when I received the transcript, I realised that this did not reflect the reality of what I had experienced during this session. I believed that it only partially represented what the PSTs had experienced. There were deletions, inaudibles, no reference to names, and the intensity of Jess' and other PSTs' emotions were missing. I transcribed the session myself, and as I did so I noticed that I was a dominant voice (Brandenburg, 2008) even though RRI sessions were designed to be a forum for PST voice and a place for them to unpack their learning; I noticed that over half of the participants in this roundtable session were silent; I noticed the intensity of Jess' frustration and anxiety. By re-examining the transcript, more was revealed about the complexity of the interactions within the roundtable discourse and this led me to more closely challenge my assumption that RRI was a conduit for PSTs to maximise their learning opportunities in the Learning and Teaching Mathematics courses.

Assumption Two: PSTs Maximise Learning Opportunities by Participating Verbally in Roundtable Reflection Inquiry Discourse

Teacher educator voice has long been privileged in teaching and learning contexts and my teacher education mathematics classrooms were no exception. Voice has been widely researched in teacher research (Connelly & Clandinin, 1990; Hamilton & Pinnegar, 1998), reflective practice (Alerby & Elidottir, 2003) and self-study (Dalmau, Hamilton, & Bodone, 2002; Berry, 2004; LaBoskey, 2004; Loughran & Russell, 2002) literatures. My self-study of mathematics learning and teaching revealed that my teacher educator *voice* had been privileged within this domain, and the ongoing reliance of PSTs on my knowledge and experience prompted me to consider alternatives, such as the creation and development of roundtable reflection as a practice. RRI was therefore designed to become a structured and formal opportunity for PSTs to unpack their learning within a socially supportive environment. An assumption underpinning RRI was that PSTs who were verbally engaged in the reflective discourse would maximise their opportunities for learning. It was imperative for me that all learners were provided with an opportunity to voice an opinion, to verbally engage. While some PSTs were willing to contribute to the discussions regarding their mathematical experiences in schools, others remained silent. By interacting verbally, teacher educators make judgements about learners, the knowledge being constructed and hence, the learning. Therefore, when teaching, a lack of (verbal) participation and interaction became significant for me as I believed that it consequently minimised the exposure of the learner's understanding and therefore, that which could be measured, judged and acknowledged. PSTs who were verbally engaged in the reflective discourse would be maximising their opportunities for learning, therefore providing learning indicators for me as a teacher educator. As Brown and Coupland (2005) have suggested often we are seduced by the "polyphony" of talk rather than by its absence (p. 2) and in retrospect, understanding this seduction has been crucial for me in coming to know more about the complexities associated with what was said within these RRI contexts and, most importantly, what remained unsaid and why. My assumption about PSTs' learning about mathematics through oral contributions to the discourse was challenged and this led me to an examination of silence during RRI sessions.

Voice and Silence

Through my ongoing analysis of roundtables, I came to understand more about silence and how silence was being understood and actively constructed by PSTs during Learning and Teaching Mathematics RRI sessions. As Pinnegar and Hamilton (2009) suggest, self-study researchers need to actively seek to "capture the voice of ... practice" and if we want to "improve our practice and ... truly understand it, we cannot silence the voice of the other" (p. 111). In doing so, this process may surface some uncomfortable realities. My assumption about PSTs' learning through oral contributions to the discourse was challenged. Through using RRI sessions to "focus my gaze" on silence (Gordon, Holland, Lahelma, & Tolonen, 2005) I came not only to understand more about PSTs' conceptions and enactment of silence, but also to challenge my own assumptions and perceptions about how silence is both identified and understood in the teacher education classroom.

In order to specifically understand silence as it was understood in the mathematics teacher education classroom, I analysed the transcripts of two RRI sessions (August, 2003) and examined the PST-written reflections from the roundtable sessions conducted over 3 years (2002–2004). These data highlighted a number of themes that were directly related to the PSTs' experiences and explanations of silence during RRI sessions. These themes included (1) fear; (2) flow and pace of the conversation; (3) contemplation; and (4) nothing to contribute.

1. "Frightened they could be wrong": Fear

Initial analysis of the transcripts together with PSTs' post-session written reflections revealed an element of PST fear. One PST stated, *they're [PSTs] frightened they could be wrong* (Gen, R12). This sentiment was also reflected by another

PST Christie, who stated that although she found the session helpful, she was also daunted by the thought of contributing to the group discussion: *I found the session helpful and interesting but I found the group a bit daunting to talk in front of* (Post-session written reflection). Some PSTs suggested that they monitored their contributions to discussions as they were afraid that they could be wrong or that they felt daunted simply by the thought of contributing to discussions. Brookfield and Preskill (2005) identify the notion of "unwanted quiet" and suggest that "students won't talk because they're afraid of making a mistake by saying something that's considered . . . unintelligent, or awkward. This is particularly true if certain students or the teacher are models of confident loquacity" (p. 180). Although I cannot confidently comment specifically on the (perceived) loquacity of individuals within the group, I can reflect on my own. In the following interaction during Roundtable 12, it became apparent that perhaps some PSTs were afraid of me and that I represented someone who was a "textbook talker" and in being so, aroused a sense of fear and verbal inadequacy in themselves:

Jack: I'll say something in basic jargon and you'll say. . .
Robyn: Do I use big words?
Jack: For me you do, sometimes. You'll say something like it could be out of a textbook. . .
Robyn: That's interesting for me
Jack: It makes you think should I have explained it like that or perhaps in a different way

Another student suggested, *we don't want to look foolish*. The notion that PSTs are both developing and protecting an identity in the process of learning about teaching is consistent with research that suggests that some learners both protect and project a particular identity and that this projection may be self-modified according to the context (Brown & Coupland, 2005). For example, some PSTs may remain quiet as a protective mechanism; others speak but then suggest that they reveal perhaps too much of their vulnerability as a pre-service teacher. PSTs' identities are forming and reforming and they choose to either project and/or protect that identity, as seen in the multiple ways in which they interact during RRI sessions.

Fear was not a feeling that I had consciously considered, especially as I had many years of experience in teaching mathematics. Yet as I collated and considered the PST feedback I began to question – what is the identity I want to project? Do I have fears as a mathematics teacher educator? Do I over-talk when I am trying to protect my identity, when I feel vulnerable?

2. "It gets too late": Flow and Pace of the Conversation

PSTs referred to the flow and pace of the conversation and to the ways in which the combination of flow and pace in conversations can preclude verbal involvement in the conversation, as the possible contribution becomes irrelevant once the moment has passed. Lin expressed this dilemma, as she experienced it, in the following way: *it's just that every time sort of I go to put something in, the conversation . . . changes and you move away so it sort of becomes irrelevant pretty quick* (R11). The reference to pace and missing the moment is mentioned by Hannah who states that, *sometimes you get off track and then it gets too late* (R12). For these and some other PSTs, the pace of the conversation became a limiting factor when attempting to engage in verbal discussions so that by the time they were prepared to contribute, the conversation had moved onto another topic. Although this should not have been a revelation for me – someone who has been teaching in various contexts for many years – it was. On reflection, my practices over many years as a teacher educator reflected my implicit and prescriptive assumption that the majority of participants in classes at university approached discussions the way that I did and therefore could quickly and immediately make the intellectual leap from one point to another. Understanding that this was not always the case has encouraged me not only to acknowledge the need to sometimes control the pace, but also to allow for reflective time during the course of the discussion – to halt the flow, as such. In doing so, we (PSTs and teacher educator) have all become intensely aware of the impact of waiting.

3. "Sitting in the shadows": Contemplation

The third category, contemplation, reveals that some PSTs preferred to *sit in the shadows* and consider the issues being discussed at the roundtable. Jim suggested that some PSTs may need to have both physical distance and more time in order to contemplate what has been discussed during RR sessions. Jim stated: *they might not want to express [their ideas]; they want to go away and think about it* (R12). Jim's suggestions also correlated closely with the reflections provided by Justin. When reflecting on the RR session he wrote that he:

> tuned out of what was happening after this comment [during the roundtable session] because my own thoughts were focused on this remark. The roundtable format allowed me to sit in the shadows and have this time to think to myself. (Justin, R15, December, 2004)

Contemplation through "tune-out" was an important practice for Justin. In this sense then, tuning out and the opportunity to "sit in the shadows" might suggest that this inner thinking might be indicative of a deeper personal exploration of the learning associated with his experience.

James also stated that:

> I learnt in this format [RRI] – even though I did not want to contribute my thoughts. I linked the things that others suggested/talked about, to my own opinions/ideas, reasoning inside my head as to whether I agreed with what was said, or did not. Without participating verbally, I still learnt about others' opinions, and allowed these to influence my own. (Post-session written reflection, December, 2004)

4. "You just don't have anything to contribute": Nothing to Contribute

PSTs also suggested that at times, their silence was because they felt they did not have anything to contribute to a discussion. My assumption was that everyone should have something to contribute. In the following statement, Don justified his silence by stating that rather than not understanding the content or experience being discussed, he had nothing to say: *you don't always have something to talk about. You might just sit there. You can still... understand what's going on... but you just don't have anything to contribute to that class... you know you haven't had that happen to you at school* (Don, R11). Don's comment suggests that if the experience being discussed connected in some way with his own experiences, then he may be more likely to contribute.

Perhaps it would be fair to conclude that there is nothing that is particularly new or even unexpected in the PST responses about what might be happening in the silences during RRI sessions. In fact, the data might also indicate that while silence can be seen as productive, the data also provide evidence to suggest that not all silence is productive. Importantly for me however, as a result of my own curiosity and systematic inquiry, silence and the impact of what remained unsaid was identified as an aspect of learning about teaching that I had not ever consciously considered even though I had been teaching mathematics for many years in many different teaching and learning contexts. What impact did this new learning have on my pedagogy as a teacher educator?

Teacher Educator Learning

In coming to understand more about how the silences were being constructed within the RRI sessions in my mathematics teacher education classroom, it was evident that this space, rather than being interpreted primarily as an absence of talk was a rich and actively constructed space. Silence did not necessarily represent disengagement and/or disinterest. It might represent: non-vocal participation (for example, James, Roundtable 14); inability to enter the conversation due to the speed of thoughts and the flow and discussion of ideas (Roundtable 12); deliberate choice (Roundtable 12); or an acknowledgement that the PST had nothing to contribute at that point (Roundtable 11).

My learning about my pedagogy through examining silence also occurred through better understanding silence as wait-time. Loughran (1996) suggests that teacher educators need to provide appropriate time for responses from their students and PSTs, prior to the teacher educator filling the void. In this sense, wait-time is a conscious strategy to impose silence in order to elicit thoughtful responses. My understanding of wait–time in this teacher education context developed. Wait–time following questioning during RRI sessions allowed me to both explore learning and teaching issues more deeply and provide some indication for me as to the level of pre-service teacher understanding.

There was an expectation that discussion would underpin the RRI process and when talk did not readily eventuate, some PSTs filled the void with talk, which may have been the result of feeling uncomfortable in the silence. The result was talk for talk's sake, rather than being prompted by a genuine need to discuss a concern or to contribute to the conversation. The analysis of the data related to verbal interactions also indicated that I felt some discomfort within, and about, silence. This is evidenced by my ongoing questioning and by my explicit attention to silence and the dominant voices at the roundtables. Changing embedded patterns of behaviour is a slow and often painful process; acknowledgement, however, is an initial step. Learning more about myself as a learner has meant that my taken-for-granted teaching practices have been challenged. The challenges have led to a deeper understanding of PST learning and the impact that my teaching has on *our* learning.

Conclusion

My initial intention when designing RRI for cohorts of PSTs in my Learning and Teaching Mathematics classes was to provide a structured opportunity for PSTs to unpack their own mathematical experiences in a meaningful way. In this chapter, I have identified and examined two of my assumptions that influenced the structure and the conduct and the expectations associated with RRI: namely that (1) the structured reflective sessions would enable the challenging of taken-for-granted assumptions about learning and teaching mathematics, and (2) PSTs maximise learning opportunities by participating verbally in RRI discourse. PSTs did unpack their mathematical experiences; they challenged their own and others' assumptions about learning and teaching mathematics and they specifically explored silence in the mathematics classroom.

However, as Kosnik (2001) has suggested, although our teaching intention is to provide support, motivation and guidance for our PSTs, through self-study we do come to know and understand more about who we are as teacher educators and how our assumptions and practices impact not only PSTs' learning but our learning. Implementing RRI with PSTs has also been a powerful way for me to examine my pedagogy – more has been revealed about the dynamics and complexities associated with teaching and learning mathematics. Transcribing my roundtable data exposed key new learning; identifying and examining the pervasiveness of silence in RRI led to deeper understandings of the ways in which the *unsaid* is understood, interpreted and enacted. Learning in mathematics teacher education is about coming to know and self-study is a methodology that enabled me as teacher educator researcher to notice, to challenge my assumptions and to name new knowledge. Ultimately this enabled me to enact new practices in my teaching (Loughran, 2006) and understand more about the subtle and complex ways that our students teach us to learn.

References

Alerby, E., & Elidottir, J. (2003). The sounds of silence: Some remarks on the value of silence in the process of reflection in relation to teaching and learning. *Reflective Practice*, *4*(1), 42–51.

Berry, A. (2004). Self-study in teaching about teaching. In J. J. Loughran, M. L. Hamilton, V. K. LaBoskey, & T. Russell (Eds.) *International handbook of self-study of teaching and teacher education practices* (pp. 1295–1332). Dordrecht, The Netherlands: Kluwer Academic Publishers.

Berry, A. (2007). *Tensions in teaching about teaching: Understanding practice as a teacher educator*. Dordrecht, The Netherlands: Springer.

Brandenburg, R. (2004a). Roundtable reflections: (Re) defining the role of the teacher educator and the preservice teacher as co-learners. *Australian Journal of Education, 48*(2), 166–181.

Brandenburg, R. (2004b) Reflective practice as a means of identifying and challenging assumptions about learning and teaching: A self-study. In D. Tidwell, L. Fitzgerald, & M. Heston (Eds.), *Risking the journey of self-study in a diverse world. Proceedings of the Fifth International Conference on the Self-Study of Teacher Education Practices* (pp. 45–49). [Herstmonceux Castle, UK] Cedar Falls, IA: University of Northern Iowa.

Brandenburg, R. (2008). *Powerful pedagogy: Self-study of a teacher educator's practice*. Dordrecht, The Netherlands: Springer.

Brookfield, S. D. (1995). *Becoming a critically reflective teacher*. San Francisco: Jossey-Bass.

Brookfield, S.D., & Preskill, S. (2005). *Discussion as a way of teaching: Tools and techniques for democratic classrooms,* (2nd ed.) San Francisco: Jossey-Bass.

Brown, A., & Coupland, C. (2005). Graduate trainees, hegemony and resistance. *Organization Studies, 29*(7), 1049–1069.

Clarke, A., & Erickson, G. (2004). Self-study: The fifth commonplace. *Australian Journal of Education, 48*(2), 199–211.

Cochran-Smith, M., & Lytle, S. L. (2004). Practitioner inquiry, knowledge, and university culture. In J. J. Loughran, M. L. Hamilton, V. K. LaBoskey, & T. Russell (Eds.), *International handbook of self-study of teaching and teacher education practices* (pp. 601–649). The Netherlands: Kluwer Academic Publishers.

Cochran-Smith, M., & K. M. Zeichner (Eds.) (2005). *Studying teacher education: The report of the AERA panel on research and teacher education*. Mahwah, NJ: Lawrence Erlbaum Associates.

Connelly, F. M., & Clandinin, D. J. (1990). Stories of experience and narrative inquiry. *Educational Researcher*, 19(5), 2–14.

Dalmau, M. C., Hamilton, M.L., & Bodone, F. (2002). Communicating self-study within the scholarship of teacher education: Herstmonceux working group. In C. Kosnik, A. Freese, & A. P. Samaras (Eds.), *Making a difference in teacher education through self-study*. Proceedings of the Fourth International Conference on Self-study of Teacher Education Practices (pp. 59–62). [Herstmonceux Castle, UK] Toronto, ON: OISE, University of Toronto.

Gordon, T., Holland, J., Lahelma, E., & Tolonen, T. (2005). Gazing with intent: Ethnographic practice in classrooms. *Qualitative Research, 5* (1), 113–131.

Hamilton, M. L., & Pinnegar, S. (1998). The value and the promise of self-study. In M. L. Hamilton, V. LaBoskey, J. J. Loughran, S. Pinegar, & T. Russell (Eds.), *Reconceptualizing teaching practice: Self-study in teacher Education* (pp. 235–246). London: Falmer Press.

Kosnik, C. (2001). The effects of an inquiry-oriented teacher education program on a faculty member: Some critical incidents and my journey. *Reflective Practice, 2*(1), 65–80.

LaBoskey, V. K. (1994). *Development of reflective practice: A study of preservice teachers*. New York: Teachers College Press.

LaBoskey, V. K. (2004). The methodology of self-study and its theoretical underpinnings. In J. J. Loughran, M. L. Hamilton, V. K. LaBoskey, & T. Russell (Eds.), *International handbook of self-study of teaching and teacher education practices* (Vol. 2, pp. 817–869). Dordrecht, The Netherlands: Kluwer.

Lankshear, C., & Knobel, M. (2004). *A handbook for teacher research: From design to implementation*. New York: Open University Press.

Loughran, J. J. (2002). Effective reflective practice: In search of meaning in learning about teaching. *Journal of Teacher Education, 53*(1), 33–43.

Loughran, J. J. (2004). Learning through self-study: The influence of purpose, participants and context. In J. J. Loughran, M. L. Hamilton, V. K. LaBoskey, & T. Russell (Eds.), *International handbook of self-study of teaching and teacher education practices* (pp. 151–192). The Netherlands: Kluwer Academic Publishers.

Loughran, J. J. (2006). *Developing a pedagogy of teacher education: Understanding teaching and learning about teaching*. London: Routledge.
Loughran, J. J. & T. Russell (eds.) (2002). *Improving teacher education practices through self-study*. London: RoutledgeFalmer.
Pinnegar, S., & Hamilton, M. L. (2009). *Self-study of practice as a genre of qualitative research: Theory, methodology, and practice*. Dordrecht, The Netherlands: Springer.
Schuck, S. (2002). Using self-study to challenge my teaching practice in mathematics education. *Reflective Practice Journal*, *3*(3), 327–337.
Schön, D. (1983). *The reflective practitioner: How professionals think in action*. New York: Basic Books.
Schön, D. (1987). *Educating the reflective practitioner: Toward a new design for teaching and learning in the professions*. San Francisco: Jossey-Bass.
Tripp, D. (1993). *Critical incidents in teaching: Developing professional judgement*. New York: Routledge.
Woods, P. (1993). *Critical events in teaching and learning*. London: The Falmer Press.

Chapter 7
Making Sense of Students' Fractional Representations Using Critical Incidents

Nell B. Cobb

Introduction

In a recent interview with a candidate for a mathematics education position at the institution where I work, I stated that I am constantly reinventing myself as a mathematics educator. This conversation started me thinking about experiences that have shaped my work in mathematics education with prospective and practising teachers. In retrospect, I see three crucial factors: my mother who had great number sense; my many years of experience in teaching secondary mathematics students; and my experience working with Algebra Project middle-school teachers and students.

During my more formative years, I was influenced by my mother who was really good with estimation and was excellent with mental mathematics calculations. When I was in elementary school, my mother and I would grocery shop together. During these shopping trips we would play an estimation game to see which one could guess an amount closest to the total for our groceries. My mother would always win. Although, I have never been as good an estimator as my mother, I do have the ability to manipulate numbers (whole, fractions and decimals) in ways that makes it easy for me to compute. This early experience paved the way for a strong interest in mathematics and eventually became part of the incentive for my undergraduate studies in mathematics and secondary education.

I cannot look back on my experiences in secondary mathematics teaching without great alarm. I was extremely procedural in my approach to teaching and taught most classes of algebra and geometry with a teacher-centred focus. After all, I was simply practising what had been modelled for me in mathematics classrooms. As a result, I did not have any student discussions, no opportunities for students to investigate mathematical relationships, and very little consideration of students' thoughts about the topics. It now frightens me to think about my former students' limited experiences in my courses. I hope, but highly doubt, that somewhere in their

N.B. Cobb (✉)
DePaul University, Chicago, IL, USA
e-mail: ncobb@depaul.edu

S. Schuck, P. Pereira (eds.), *What Counts in Teaching Mathematics*, Self Study of Teaching and Teacher Education Practices 11, DOI 10.1007/978-94-007-0461-9_7,

mathematics studies they have had a more student-centred experience. After many years of what I felt was effective teaching, I decided to try my hand at administration.

I applied for a position advertised in a Chicago newspaper for a director of a middle-school mathematics program, and in 1992 became the first director of the Chicago Algebra Project. Although I was not sure about this move at the time, I now see this as one of the significant events in my career as a mathematics educator. Before then, basic mathematics concepts were things that I took for granted, never really giving them much thought. My mother had prepared me to think intuitively about what I now know as natural and whole numbers, and when I taught mathematics in secondary school I did not have to confront the difficulties that elementary teachers face. It was only when I started to work with middle school teachers in the Algebra Project that I began to realize the complexities underlying apparently simple ideas. When I then went on to teach mathematics for elementary teachers at the university level, it became clear how much the prospective teachers needed to learn about elementary mathematics concepts. More importantly, I came to realize how much work is needed to help them deepen their understanding of numbers, especially rational numbers. My growing understanding about this lack of mathematical preparation – mine as well as theirs – is the subject of this chapter.

The Algebra Project

The Algebra Project is a mathematics education movement that has existed in the United States for over 25 years. The founder of the Algebra Project, Robert (Bob) P. Moses, was captivated and motivated by the "sit-in" movement of the south. This movement was organized by black youth to address the injustice in America over 50 years ago. Bob became a leading figure in the Civil Rights Movement of the 1960s, and part of the drive for poor black sharecroppers to gain the right to vote in Mississippi. In a recent article in the *Harvard Educational Review*, Bob describes the framework for this change as a "Pulling from the Top, pushing from the Bottom" movement. People in administrative positions, including the President and Congress, acknowledged the forceful voice of people who understood the energy needed to put change in motion (Moses, 2009).

The same organizing strategy used in the Civil Rights Movement guided the beginning of the Algebra Project. The project was originally a classroom intervention, then grew into a middle school program, and is currently a high school program for cohorts of students. The Project has always been based on the premise that algebra, the gatekeeper for higher mathematics learning, is a civil right for all students. The program is designed to increase the number of students underrepresented in advanced high school mathematics courses and also increase the number of students successfully completing college level mathematics courses. More importantly, the Algebra Project advocates that the change needed to successfully prepare underrepresented students for the twenty-first century must be led by representatives of that population (Moses, 2001).

As director of the Chicago Algebra Project, I was charged with the responsibility of providing leadership for a staff of three people and a small group of Lead Teachers. The staff was responsible for the day-to-day work of the project in schools and their respective communities. The Algebra Project Lead Teachers were responsible for providing instructional support to middle school teachers on the Algebra Project Transition Curriculum. This material was designed to prepare middle school students to successfully complete high school algebra. In my leadership role I also worked with the Lead Teachers to: model Algebra Project curriculum implementation; collaborate with teachers to identify implementation problems and solutions to those problems; and conduct demonstration lessons for the broader Chicago community.

The Algebra Project has developed a Model of Excellence in mathematics teaching that describes the basic competencies that mathematics educators possess or need to possess. Two of these competencies – accurate empathy and intellectual flexibility – are the focus of my self-study.

> Accurate empathy is the ability to listen to and understand the thinking, feeling, and behavior of others; and appropriately respond in an effective manner. Understanding can be captured from both verbal and non-verbal communication of others (i.e., facial expressions, body language, eye contact, and lack of participation). The response can be demonstrated by paraphrasing the words of others and "checking in" around how a person's feeling about things in an effort to improve one's understanding of the situation (Algebra Project, 2009, p. 1).
>
> Intellectual flexibility is the ability to make connections among disparate ideas and experiences, including mathematics concepts, everyday life and other subject areas (Algebra Project, 1999, p. 24).

The Context

After using five different editions of a textbook on mathematics for elementary teachers, I was reluctant to switch to a new text. I remember thinking that the new textbook could not be that much different from the one we already used. I was a bit cynical about the whole textbook switch. However, when I took the time to read the text thoroughly, plan lessons and eventually use the new text and the manual on accompanying activities, I found a significant difference in how mathematics concepts were presented, and how prospective teachers were given opportunities to develop new meaning for these concepts. The new text presents multiple learning experiences for prospective teachers to extensively and routinely investigate and revisit their understanding of number. I finally listened to my colleagues in the mathematics department and realized that this new text was designed to have prospective teachers investigate mathematics concepts in a variety of conceptual and procedural ways.

Students discussed in this chapter are either prospective elementary teachers or practising middle and high school teachers. The prospective teachers are required to take two mathematics content courses for elementary teachers followed by a course in mathematics education. Prospective teachers can determine at the end of each course whether they will continue with the same instructor or choose to enroll in another section of the course with a different instructor. Over the past decade, approximately 70% of my students have remained as an intact group in any given year. Thus, for the most part students form a cohort studying mathematics teaching and learning for one academic year with me. Most prospective teachers in this sequence struggle with procedural and conceptual mathematics learning. Many of these students wait until the last part of their studies to take these courses because they have had bad mathematics learning experiences at the elementary and secondary levels. As well, there are prospective teachers who have strong mathematical procedural knowledge, but they sometimes struggle and are resistant to gaining appropriate conceptual knowledge.

The middle school teachers are part of a new Master's degree program designed jointly by the mathematics department and the School of Education. This program, which leads to an endorsement in middle school mathematics, addresses both mathematics content and pedagogy in each course. These teachers move through the program as a cohort and are required to take nine courses for their mathematics endorsement with the option of taking three additional courses to complete a Master's degree.

The secondary teachers discussed in this chapter were a part of a summer two-week Algebra Project Professional Development Institute. The purpose of the Institute was to develop strategies to bridge the multiple cultures of university and K-12 practitioners and work more effectively with teachers in institutes, workshops and schools. I was a co-facilitator for this Institute. This effort was funded by the National Science Foundation's Discovery Research K-12 initiative.

In the mathematics courses for both undergraduate and graduate programs, students are encouraged to develop a deep understanding of elementary mathematics concepts. These courses are activity-based, allowing students to think deeply about what they already know and need to learn about elementary school mathematics. The courses focus on helping prospective and practising teachers enhance their mathematical proficiencies. As outlined by the National Research Council, these are conceptual understanding, procedural fluency, strategic competence, adaptive reasoning, and productive disposition (Kilpatrick, Swafford, & Findell, 2001). "Effective programs of teacher preparation and professional development help teachers understand the mathematics they teach, how their students learn that mathematics, and how to facilitate that learning" (p. 10). The specific mathematics topics in these programs include: problem solving, the decimal system and place value, and fractions. Prospective and practising teachers explore the interpretations of addition, subtraction, multiplication and division of all numbers.

The undergraduate and graduate mathematics programs provide a unique self-study opportunity for me to learn from both populations. Loughran (2005) states that by engaging in self-study teacher educators are able to examine the complexities of teaching in an effort to gain a deeper understanding of the practice. By engaging in self-study, I hope to enhance prospective and practising teachers' learning and consequently impact elementary student learning.

Critical Incidents

The study of my teaching and learning starts with some exemplars of critical incidents that describe times when I was most and least effective as a mathematics educator. Reflecting on my practice using critical incidents is an excellent self-study tool because it helps me capture stories from my experiences while working with prospective and practising teachers. These stories address six key questions: what was the situation; what events led to this situation; who was involved; what did I think, feel, or want to do in the situation; what did I actually do or say; and what was the outcome? It should be noted that not every question is answered in each incident and the incidents are actually problem solving opportunities.

This self-reflection process is described in Spencer and Spencer (1993) as Flanagan's Critical Incident Method. It is an interviewing protocol that allows the interviewee to identify and discuss the most critical situations that they encounter in their jobs. Spencer and Spencer took this interview process one step further and developed another interview strategy, the Behavioral Events Interview (BEI). The BEI process goes beyond the Critical Incident Method to identify competencies needed to do the job well. Many companies and educational organizations use the BEIs to develop competency models that focus on superior performance in the work place. The Algebra Project's Model of Excellence in mathematics teaching uses the BEI interviews and coding process as developed by the Hay Group (Cobb, 2001).

I have used the Critical Incident Method to capture stories involving various participants' investigation of fractional representations. I will review reflective scenarios collected over time to see how I can strengthen my responses to students by displaying more accurate empathy and intellectual flexibility in an effort to understand and enhance students' thinking and use of fractional representations. Here are the stories:

Incident #1: What Is the Whole?

In this scenario, my intention was to encourage prospective teachers to take particular care when defining the unit or "the whole" associated with a fraction. I assigned an article for the class to read before we did any collective work with fractions. In the article there was one example that I had students respond to:

> In an effort to use constructivist mathematics methods that include concrete, hands-on representation of mathematical concepts, many school teachers use egg cartons to represent fractional values. The fraction 1/2 can readily be shown by covering six sections of the egg carton, 1/3 by covering four sections, and 1/4 by covering three sections. A first year teacher, Mary, was asked to represent 1/5 [of an egg carton]. She immediately cut a five-section piece and a one-section piece from the carton, and then covered one section of the five-section piece with the one-section piece (Wentworth & Monroe, 1995, p. 357).

Here is the figure:

I thought that this reading would be a good precursor to the work that prospective teachers would do in their groups. So I asked them to respond to two questions:

1. Why is defining the unit or "the whole" in fraction examples often confusing to elementary students?
2. In the example about Mary, what were the mistakes that were made by this first year teacher?

This group of prospective teachers had been working on activities involving numbers and the decimal system, including representing numbers with bundled objects and misconceptions in comparing decimal numbers. I felt that this group on the whole was fairly comfortable with these decimal system topics as demonstrated by their class work and related discussions, homework and assessments. It was time to move on to investigate the meaning of fractions.

Because I focused on the particular example, it was evident that most only read that part of the article even though I had assigned the whole article. The responses were mostly taken out of the article, in some cases verbatim, with not much thought about the significance of defining the whole. Mary's mistake of changing the whole or unit associated with an egg carton was explained by many as an error that most people should not make. However, I have seen this error made several times by more experienced teachers. In addition, I was not convinced that these prospective teachers could concretely determine 1/5 of an egg carton. It is important to note that, at this point, the idea of taking a fraction of a whole item was the only thing I considered.

I had expected students to read the complete article even though the rest of the article switched to computation involving student representations of multiplication and division of fractions. At this point, I did not want to have the discussion about these computations until we investigated multiplication and division of fractions in the activities manual. In the next class, I asked the prospective teachers to consider what taking 1/5 of the contents of an egg carton would mean and suggested that they postpone discussions about other parts of the article until later. They were asked to draw the scenario. This took a while, but eventually students could explain why using 5 sections of an egg carton and taking 1/5 of that section was much easier than, but not appropriate for, taking 1/5 of the carton which contains 12 sections. It was important for the class to talk about when this scenario could realistically happen. Students went back to the definition of a fraction, which in this case they said was part of an object. Their question was, "Why would you take 1/5 of a dozen of eggs or 1/5 of an egg carton?" Students concluded that 1/5 of an egg carton could happen, but they didn't understand when or why this would happen. I left this inquiry open for possible discussion at another time but realized, after the fact, that this article (more specifically the example) really didn't clarify the concept but rather raised more questions.

I continued to use this article for the next class and had students discuss another scenario. The authors went back to the egg carton representation and asked this teacher to also show 1/3 of the egg carton. The teacher used a three-section piece and used the same one-section piece to cover one part of the three-section piece. The teacher self-corrected by declaring that she had changed the unit and recognized that the unit must remain constant; the 12-section egg carton should have been the unit for this problem.

Another example that the authors (Wentworth & Monroe, 1995) use comes from Maher and Davis (1990). In this example, elementary students demonstrated an understanding of the whole when using pizza pieces. The problem involved two pizzas, each cut into twelve pieces. "If seven students ate one piece from each pizza, what fraction of the two pizzas was eaten?" (p. 68). One student decided 14/24 pieces were eaten whereas the teacher answered 14/12 pieces. This teacher changed her answer on reflection as she determined that students would have eaten more pizza than they possessed if the 14/12 represented the total eaten.

In both teacher-centred scenarios, the teacher self-corrected. The impact of teacher knowledge on student achievement is well documented (Hill, Schilling, & Ball, 2004; Hill, Rowan, & Ball, 2005; Ma, 1999). The Study of Instructional Improvement discussed in Hill et al. (2005), which involved around 3,000 first and third grade students and 700 of their teachers, found that teachers' performance on the knowledge for teaching questions was a strong indicator for student achievement. The better the teacher scored on the Content Knowledge for Teaching Mathematics (CRT-M) items, the better their students achieved in a school year as measured by the Terra Nova assessment (Hill et al., 2005).

It was important for me to get the prospective teachers' reaction to these scenarios because there will be times when their responses to elementary students might not be correct. However, these prospective teachers agreed that understanding the question, "What is the whole or unit associated with the problem?" would clarify their thinking. They are not very comfortable in thinking about not knowing answers or making mistakes in front of a group of students, but they are comfortable striving to understand student thinking and responding appropriately. This is a good first step and one that I try hard to model.

Incident #2: Can These Stories Be Represented by the Same Expression?

In this scenario, prospective teachers worked in small groups to answer the following:

> For the following story problem, determine whether the problem can be solved by subtracting 1/2 − 1/3. If not, explain why not, and explain how the problem should be solved if there is enough information to do so. If there is not enough information to solve the problem, explain why not.
>
> (a) Zelha pours 1/2 cup of water into an empty bowl. Then Zelha pours out 1/3 cup of water. How much water is in the bowl now? (Beckmann, 2008a, p. 90).

Over the years, I have collected prospective teachers' responses to this and other items to determine the effectiveness of my teaching. The first time students see this problem is when they are working on finding a common denominator using pictures. They are encouraged to draw a picture of the situation, but few do so. It is worth remembering, though, that they are only in the fifth week of the first content course, and they still see this problem through a procedural lens. Thus, students find the common denominator, change the fractions to equivalent fractions and subtract. Instead of a picture they write on the side of the paper:

$$1/2 = 3/6$$
$$1/3 = 2/6$$
$$3/6 - 2/6 = 1/6$$

When this happens, I really want to yell "don't do that!" But I restrain myself and ask, "why do you have this work on the side of your paper?" In most cases the answer is that this is the only way that they know how to get to a picture. I check my understanding of their thinking by asking questions. Must the symbols always drive the picture that you draw? How could you rethink this problem or a similar problem conceptually? Where or when in real life would you find a problem like this to use as a context?

To get a further sense of prospective teachers' thinking, I chose two other problems from the same set to help students develop the skill of having the picture drive the symbolic representation. These problems use a similar situation but present totally different contexts:

> (b) Zelha pours 1/2 cup of water into an empty bowl. Then Zelha pours out 1/3. How much water is in the bowl now?
> (c) Zelha pours 1/2 cup of water into an empty bowl. Then Zelha pours out 1/3 of the water that is in the bowl. How much water is in the bowl now? (Beckmann, 2008a, p. 90)

Taking these questions as a group, prospective teachers consider the similarities and differences within this group. The questions I try to answer at this point are whether or not my students can identify the whole in each scenario and if they are thinking that the whole changes or remains the same in each problem. I ask myself, how are students drawing pictures to represent the scenarios, are they always using the same model (for example, rectangular, large tubes, or other pictures)? In most cases the answer is "yes". These students often use rectangular shaped models because that is the way I generally model the problems. I had a few students who drew a picture of bowls.

For problem (b), many prospective teachers state that $1/2 - 1/3$ does represent this scenario. When asked why they think that, my students state that you place 1/2 cup water in the bowl and you take out 1/3 cup of water. They sometimes think that the cup is understood in the statement, "Then Zelha pours out 1/3". To help clarify the unit and to generate concrete examples to illustrate abstract concepts, I ask students to think of other scenarios that we could come up with that will illustrate that it is not always the case that "the 1/3 poured out" would be measured in terms of a cup. Students come up with 1/3 of the contents of the bowl, which is the same as problem (c), or 1/3 of some other unit that you can name, like a tablespoon. I take whatever answers are presented.

What is interesting about problem (c) is that by the time we have the conversations above, this problem naturally follows the others. Here students tend to notice that the whole in the bowl is 1/2 cup and they are taking out 1/3 of the 1/2 cup of water or 1/6 of a cup of water. So we go from 1 cup being the whole for the initial problem to 1/2 cup as the whole unit in the final part of the problem. Changing the unit is not an easy task for many prospective teachers.

Incident #3: If the Whole Keeps Changing, What Really Is the whole?

One really useful aspect of the materials used in the Beckmann text is that they enable us to keep revisiting the concept of the whole unit. In the following examples, prospective students were getting confused with determining the initial whole unit and then changing the unit to appropriately fit the problem. The problem:

> You want to make a recipe that calls for 2/3 of a cup of flour, but you only have 1/2 of a cup of flour left. Assuming you have enough of the other ingredients, what fraction of the recipe can you make?
>
> (a) Draw pictures to help you solve this problem. Explain why your answer is correct.
>
> (b) In solving the problem, how do 2/3 and 1/2 appear in different forms?
>
> (c) What are the different wholes associated with the fractions in this problem? In other words, for each fraction in this problem, and its solution, what is it the fraction of? (Beckmann, 2008b, p. 95)

I knew that this multi-step problem would not be an easy one for my students to follow, so I started a conversation around strategies for solving the problem. We simply talked through various ways to approach the problem to get to the significant question: when you take a fraction of a fraction will the result be more or less than the original recipe? In this discussion, students expressed the opinion that the recipe could be either more or less by taking a fraction of a fraction. This was an important discussion to have because it is generally assumed that when you multiply numbers the result is always bigger. Graeber, Tirosh, and Glover (1989) found these misconceptions with pre-service and elementary teachers over 20 years ago. Tobias (2009) points out that these misconceptions still exist. She states that:

> Two ideas became taken-as-shared with fraction multiplication. The first was that multiplication can be represented as "a groups of" situation or taking a fraction of a fraction. The second was the idea of distributing when the number of groups is greater than one. Though the class could solve each multiplication problem with a procedure, they never got to a point where they demonstrated conceptual understanding of the procedure for multiplying fractions. (p. 187).

However, my students appeared to demonstrate conceptual understanding. When answering questions (a), (b) and (c) above they provided the following explanations:

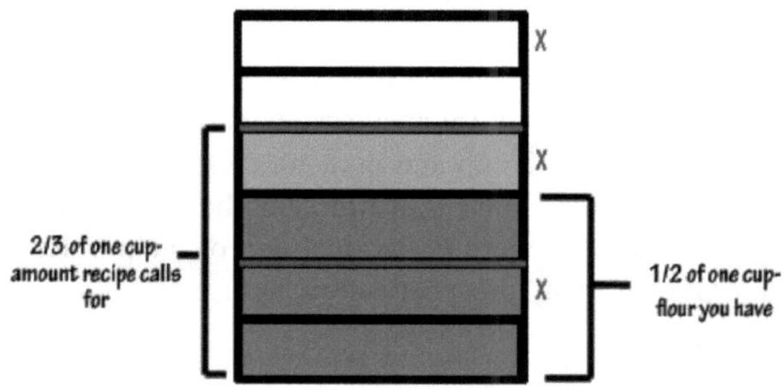

First I drew out the whole cup. Then I divided the whole cup into 3 equal parts and shaded the 2/3 cup that the recipe calls for. Since 1/2 of one cup does not have the same equal parts to compare to 2/3 of a cup (different denominators), I divided each of the thirds in half. I did this step because you can think of shading 1/2 of each third in the cup to find how many sixths are shaded by using 1/2 cup. By doing this I found that 3/6 is shaded. Then to find what fraction of the recipe you are able to cook, I looked at 2/3 as the whole and the amount eaten was 3/4 of the recipe.

$$2/3 \text{ becomes:} \frac{2 \times 2}{3 \times 2} = \frac{4}{6} \text{ and } 1/2 \text{ becomes } \frac{1 \times 3}{2 \times 3} = \frac{3}{6}$$

Part	Whole
2/3 cup	1 cup
½ cup	1 cup
¾ cup	2/3cup

Another student did a similar representation and explanation for a second assigned problem with a different context:

> Two-thirds of a cup of Healthy SnackOs provides your full daily value of vitamin B. You ate 1/2 of a cup of Healthy SnackOs. What fraction of your daily value of vitamin B did you get in the Healthy SnackOs you ate? (Beckman, 2008b, p. 95)

The prospective teacher wrote: First I drew one whole cup, then I divided it into thirds. I shaded the 2/3 cup that is the daily value of vitamin B. Next, I looked at the whole cup, and marked the halfway point. I shaded the 1/2 cup that was eaten. Then, since 2/3 and 1/2 had different denominators, I looked at the thirds and broke them

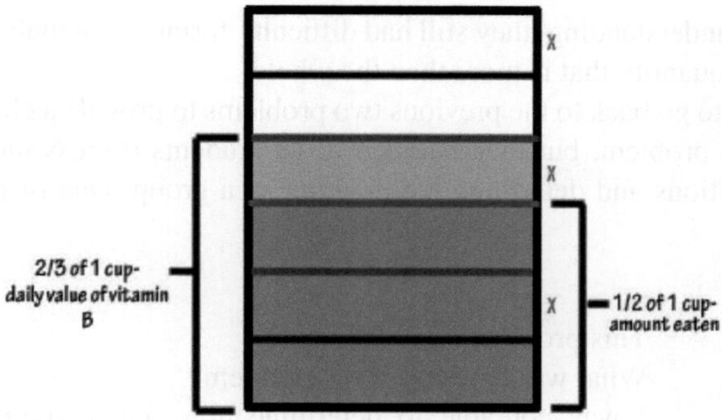

into two separate parts because if you had taken 1/2 of each third you would get the same total (3 shaded sixths). To find what fraction of the daily value of vitamin B was eaten, you have to look at the 2/3 shaded as the whole. Therefore, the 1/2 cup eaten is 3/4 of the daily value of vitamin B.

$$2/3 \text{ becomes: } \frac{2 \times 2}{3 \times 2} = \frac{4}{6} \text{ and } 1/2 \text{ becomes } \frac{1 \times 3}{2 \times 3} = \frac{3}{6}$$

Part	Whole
2/3 cup	1 cup
½ cup	1 cup
¾ cup	2/3cup

After similar drawings and explanations were consistently made by most prospective teachers, I concluded that they had demonstrated conceptual understanding of the procedure. But did they? This year I provided an assessment item that actually asked what turned out to be a difficult question for most students to answer. Students had previously stated that you could take a fraction of a fraction and get a result that is more than the original fraction. I asked:

> Two thirds of a cup of Healthy SnackOs provides your full daily value of vitamin B. You ate one cup of Healthy SnackOs. What fraction of your daily value of vitamin B did you get in the Healthy SnackOs you ate?

This simple revision revealed their lack of understanding of what they were trying to determine two thirds of. Most of my students in three subsequent cohorts had no idea what to do with this problem. Although I thought these pre-service teachers had demonstrated understanding, they still had difficulty formulating their knowledge of a fraction of a quantity that is more than the whole.

I was about to go back to the previous two problems to provide a clue for students to address this problem, but then decided to let students discuss their challenges and misconceptions and determine the drawing as a group. One of the discussion records:

Student (1):	This problem was really hard.
Teacher:	What was hard about the problem?
Student (2):	I was not able to determine how much vitamin D was in one cup.
Teacher:	Was that the problem for most of you?
Multiple students:	Yes. (Also indicated with headshakes.)
Teacher:	How do we relate this problem to the one you worked for homework? The problem was similar except in that problem, you ate 1/2 cup instead of 1 cup of Healthy SnackOs.
Student (3):	I actually solved that problem again and doubled the result. Only I could not decide on how to draw the picture.
Teacher:	What about that solution? Do we agree that you can get the result by doubling the answer from the previous problem which had you eating 1/2 cup of Healthy SnackOs?
Student (2):	That was a good idea. However, I really want to know how you would draw the picture.

Students, working in small groups to determine an appropriate drawing, eventually came up with a drawing similar to the ones previously presented. In this case, working in groups resulted in their greater conceptual understanding.

Incident #4: Working with Equivalent Fractions and Simplifying Fractions

Here is another story involving prospective elementary teachers but this time they are trying to figure out how to picture the procedure for reducing a fraction to lowest terms. The text presented the following problem:

$$6/15 = 2*3/\ 5*3 = 2/5$$

(Beckman, 2008a, p. 48)
Beckman then displayed the following color-coded model which I recreated using card stock paper to make the squares and rectangles.

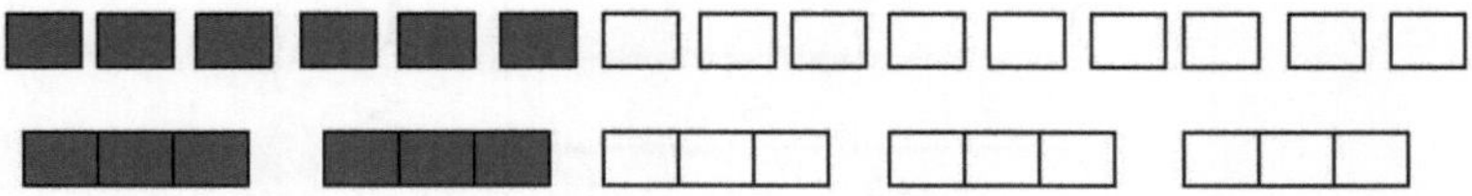

Because they also wanted to represent the 2/5, the prospective teachers added the following line:

Last year, one prospective teacher asked, "If we start with 6/15 shaded, what would happen if my future students grouped only two sections together on the second line?" When this happened, I thought that this was a good effort to anticipate student responses to a diagram. This question led to a conversation about what exactly is happening here. If students come up with groupings that work, they would have to eventually get to the same representation as line (2), three small cubes are placed together to make equal groups with 3 in each. This is represented by the multiplication problem, (2*3)/(5*3). After further discussion with the prospective teacher, she actually turned the question around and asked what would happen if we started with the last line first. It was determined that the second line in the drawing could be represented by infinitely many representations. This meant that the 2/5 could be divided into 2 parts each, 3 parts each (which we already have), 5 parts each and so on. The equivalent fractions would be 4/10, 6/15, and 10/25, respectively.

I was feeling good about the extension of the discussion because now students seemed to have a mental image for seeing equivalent representations of the same fraction. David and Meyer (1990) stated, "In general we mean primarily a mental representation, although it often happens that one makes use of paper and pencil, or even of physical materials, to help out in the process of building an adequate representation" (p. 65). The students discussed in this incident had slightly different mental images of the original diagram and adjusted the diagram to represent their thinking. In the equivalent and simplifying fractions discussion, I tried to provide an environment for students to make conjectures, share their thinking, and eventually come up with new representations.

Incident # 5: Vectors Show Fraction Division

During a summer professional development workshop for secondary Algebra Project teachers, a discussion about remediation for secondary students surfaced. The teachers stated that secondary students are still struggling with fractional representations and computation. After much discussion about student remediation, I asked teachers to consider how they could represent (1/2) ÷ (1/4). Many of the participants represented the problem using either the "invert and multiply" process, or drew a picture using an array model or showed the numbers on a number line. However, I did not expect the following illustration and explanation:

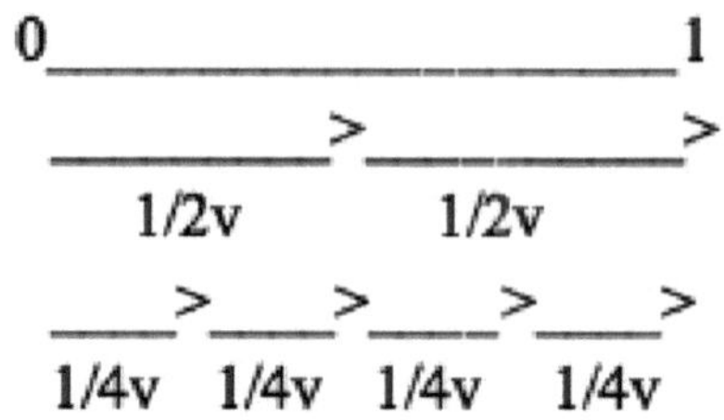

I chose to use a vector drawing to explain 1/2 divided by 1/4 by imagining a trip from 0 to 1/2. The vector length I wanted to use was 1/4 v. The question that I wanted to answer was; how many 1/4 units would you need to travel a distance of 1/2 units?

Looking at the vector drawing, I could see that it takes two vectors of size 1/4 v to travel the same distance as 1/2 units. If $2(1/4\ v) = 1/2\ v$, then $1/2\ v/1/4\ v = 2(1/4\ v)/1/4\ v = 2$.

This teacher related the division problem to the Algebra Project Modules used with his students the previous academic year. There are two modules *Trip Line* and *Race Against Time* in which students explore math concepts of number computation and vector representations. In *Trip Line*, students explore number as an object and location on a trip line and later a number line. The focus of the *Trip Line* is the class trip taken at the beginning of the module and later used to construct their trip line and then a real number line. This module is used to help students discuss and explore numbers, computation, and rules of arithmetic. The second module *Race Against Time* extends the discussion of the first module using various math concepts including vector representation.

I thought that this connection the teacher made between the work that students did with the Algebra Project materials and remediation had been overlooked and that it would be an opportunity to develop supplemental remediation materials for teachers. This teacher had provided a perfect example of integrating remediation and grade level instruction. Since I am a member of the Teacher Resource Team, I would need to more closely focus on bridging high school students' conceptual understanding gaps within the context of Algebra Project materials. This goal would help to extend my intellectual flexibility around remediation by making links between grade level instruction and remediation.

Incident #6: What Is a Double Number Line? Did It Get Introduced Too Soon?

This scenario comes from the second course that the middle school teachers took during a summer session that focused on mathematical thinking. I also used the Beckmann textbook and activity manual. The following problem was investigated by students:

Tonya and Chrissy are trying to understand 1÷ 2/3 by using the following story problem: One serving of rice is 2/3 of a cup. I ate 1 cup of rice. How many servings of rice did I eat?

To solve the problem, Tonya and Chrissy draw a diagram of a large cup divided into three equal pieces, and then shade two of those pieces.

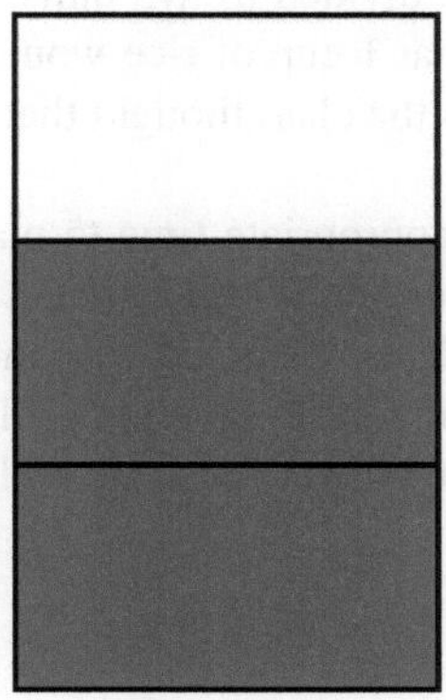

Tonya says, "There is one 2/3 cup serving of rice in 1 cup and there is 1/3 cup of rice left over, so the answer should be 1 1/3". Chrissy says, "The part left over is 1/3 cup of rice, but the answer is supposed to be 3/2 = 1 1/2. Did we do something wrong?" Help Tonya and Chrissy. (Beckmann, 2008a, p. 212).

Most groups addressed the problem of Tonya and Chrissy by discussing the appropriate whole in the answer. All groups except one either recreated the given drawing, discussed the misconception that Tonya and Chrissy had, or talked about the kinds of questions of clarification and elaboration that teachers could ask the girls. The questions were, "What was the whole associated with 2/3 of a cup?" and "What is the whole associated with the 1/3 cup of rice that was left over?" As I walked around the different groups and saw the different diagrams and suggestions for helping students self-correct, I saw that one member of a group used a double number line to illustrate this problem. This representation was introduced in a section later in the book. My question to myself was, "how was I going to spend time to explain the use of the double number line at the point when I only wanted a discussion of the model presented by the problem?" A more advanced level of Intellectual Flexibility indicates that a teacher should adapt new processes, activities or teaching strategies to the needs of the particular group (Algebra Project, 1999, p. 24). So I had the group member discuss how they used the double number line to explain the problem. Here is what the student represented:

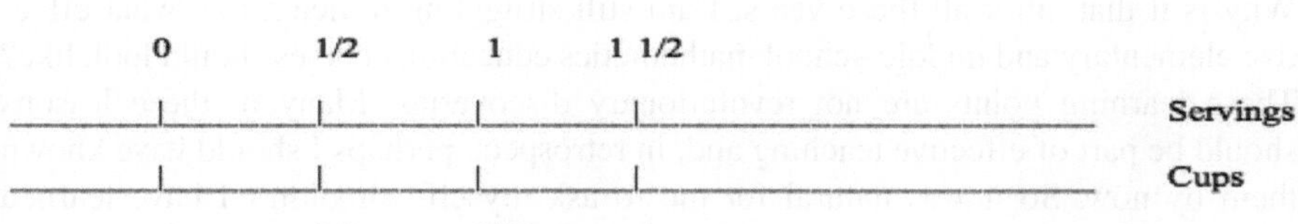

The student explained that he was introduced to the double number line in another course. He thought this was another illustration that Tonya and Chrissy could consider. In this way, they could see that 1 serving was represented by 2/3 of a cup of rice and that the servings in this number line were divided by 1/2 units, in the same way, the cups were divided by 1/3 units. If this is the case, it would be very easy for the girls to see that 1 cup of rice would represent 1 1/2 serving. The other middle school teachers in the class thought that this representation provided a much clearer representation.

I decided that this was the appropriate time to make the connection to the section on "Using Double Number Lines" and after a short break, went directly to discuss that section. Many students struggled with the difference in scales between the two number lines. After completing several problems and much discussion, the confusion about scaling became clearer for most students.

Conclusion

There were three basic learning points that I discovered or rediscovered while working with pre-service elementary, practising middle school and high school mathematics teachers.

- The use of articles discussing misconceptions might lead to more student confusion about the math topic. The first incident, "What is the Whole", demonstrates this point.
- Asking teachers to embark on the task of identifying the initial whole in a problem, discuss the associated whole for each fraction of a problem, and recognize when the initial whole has changed is not a trivial task. Incidents 2 and 3 are focused on this learning point. These stories demonstrate the challenges of getting students to understand the task and its importance.
- Sometimes I really need to listen more attentively and share classroom facilitation to help students investigate and extend their learning. In incident 4, "Working with equivalent fractions and simplifying fractions", a pre-service teacher decided to flip the order of the problem given in the text to make sense of various forms of equivalent fractions. In scenario 5, a secondary teacher provided an example of remediation within the context of grade level instruction. Incident 6 is taken from my experience with middle school teachers' use of new tools to represent fractions and proportional reasoning.

Why is it that, after all these years, I am still struggling to determine what effective elementary and middle-school mathematics education courses should look like? These learning points are not revolutionary discoveries. Many of these lessons should be part of effective teaching and, in retrospect, perhaps I should have known them by now. So it was natural for me to ask myself "shouldn't I have learned lessons things by now?" My answer is, "yes you have learned several lessons and you must continue to learn more". Teaching mathematics teachers, like teaching in general, requires that you evolve and retool periodically. I think about teaching very differently now than I did several years ago and even last year. I am in a unique position to collaborate with colleagues in the mathematics department, school of education, and the Algebra Project. It is my hope that we will be able to form a mathematics teaching and learning community and develop a structure in which we discuss different approaches to various mathematics content and pedagogy. This collaboration will also help me address and use the lessons learned from the analysis of the critical incidents.

Some short-term goals that I have identified are a need to do a better job of assessing what, when and how to use mathematics related articles, cases and other resources demonstrating misunderstandings around mathematics concepts. When these resources are used inappropriately students are left asking questions like, "So what's the point?" "Why is this important?" and "Is this a realistic problem to ask people to solve?" As an instructor I should consider more extensively the question, "In what ways will this reading increase student understanding and in what ways will it create more confusion about the mathematics concept or an instructional strategy?"

The power of prospective and practising teachers "doing" mathematics, sharing multiple strategies, and considering various model representations can address the challenge of getting teachers to understand the importance of a mathematics task. I have identified two competencies, accurate empathy and intellectual flexibility, as areas that could significantly strengthen my delivery of instruction. In scenarios 2–6, the behaviour indicators for intellectual flexibility were mostly identified. I should work to generate concrete examples to illustrate abstract concepts, use specific participant data to support my assertions, and use analogies or metaphors to clarify a concept or principal. In incident 1, I used Accurate Empathy by acknowledging participants' insights and opinions, identifying primary concerns of others, and referencing specific comment or behaviours of participants to illustrate a point.

The book *Adding It Up* concludes that effective professional development and pre-service education should be designed to help teachers on every level understand the content, determine how students learn, and provide support for effective instruction (Kilpatrick et al., 2001). As I strengthen my teaching competencies, I will address these goals. There is much work to do to attain them but the rewards for the work are enormous.

References

Beckmann, S. (2008a). *Activities manual to accompany mathematics for elementary teachers*. Boston: Pearson Education.

Beckmann, S. (2008b). *Mathematics for elementary teachers. Boston: Pearson Education*. Boston: Pearson Education.

Cobb, N. (2001). The Algebra Project model of excellence: What outstanding Algebra Project teachers and trainers look like. *Journal of Mathematics Education Leadership*, *5*(1), 33–44.

Graeber, A. T., Tirosh, D., & Glover, R. (1989). Preservice teachers misconception in solving verbal problems in multiplication and division. *Journal for Research in Mathematics Education, 20*, 95–102.

Hill, H. C., Rowan, B., & Ball, D. L. (2005). Effects of teachers' mathematical knowledge for teaching on student achievement. *American Educational Research Journal, 42*(2), 371–406.

Hill, H. C., Schilling, S. G., & Ball, D. L. (2004). Developing measures of teachers' mathematics knowledge for teaching. *Elementary School Journal, 105*(1), 11–30.

Kilpatrick, J., Swafford, J., & Findell, B. (Eds.) (2001). *Adding it up: Helping children learn mathematics*. Washington, DC: National Academy Press.

Loughran, J. (2005). Researching teaching about teaching: Self-study of teacher education. In J. J. Loughran & T. Russell (Eds.), *Studying Teacher Education: A Journal of Self-Study of Teacher Education Practices, 1*(1), 5–16.

Ma, L. (1999). *Knowing and teaching elementary mathematics: Teachers' understandings of fundamental mathematics in China and the United States*. Mahwah, NJ: Lawrence Erlbaum.

Maher, C. A., & Davis, R. B. (1990). Building representations of children's meanings. *Journal for Research in Mathematics Education*, Monograph Series (4), 79–210.

Moses, R. P. (2009). An earned insurgency: Quality education as a constitutional right. *Harvard Educational Review, 72*(2), 370–381.

Moses, R. P., & Cobb, C. E. (2001). *Radical equations: Civil rights from Mississippi to the Algebra Project*. Boston: Beacon Press.

Project, A. (1999). *Algebra Project model of excellence*. Cambridge, MA: Algebra Project Inc.

Project, A. (2009). *The Algebra Project competencies*. Cambridge, MA: unpublished.

Spencer, L. S., & Spencer, S. M. (1993). *Competence at work: Models for superior performance*. New York: Wiley.

Tobias, J. M. (2009). *Preservice elementary teachers' development of rational number understanding through the social perspective and the relationship among social and individual environments*. Unpublished Doctoral Dissertation. University of Central Florida.

Wentworth, N., & Monroe, E. E. (1995). What is the whole? *Mathematics Teaching in the Middle School, 1*(5), 356–360.

Chapter 8
Reforming Mathematics Teacher Education Through Self-Study

Joanne E. Goodell

Introduction

I have been teaching secondary mathematics teaching methods classes since the mid 1990s during which time the push for reform in mathematics education has become ever stronger. One of the major obstacles both pre-service and new teachers face in learning to teach is not being able to implement reformed (constructivist) teaching practices in a supportive environment. Pre-service teachers feel obliged to follow the goals and use the methods modeled by their mentor teacher, which may or may not match reform goals. New teachers too often find themselves in departments where the majority of established teachers do not embrace constructivism as a theory of learning. Consequently they use very traditional expository pedagogies. Therefore, my overarching goals for my students in the one mathematics methods course they take is to have them experience a constructivist learning environment, to implement constructivist teaching practices in their own teaching, and to reflect in writing and through class discussion about the effectiveness and implications for student learning of this type of pedagogy. For many, this is the first experience of a constructivist learning environment. Developing the habits of mind to learn from their reflections is the mechanism by which I seek to influence their growth as teachers beyond their contact with me in a class that meets once a week for 4 hours over the course of a semester (henceforth referred to as a "4-semester-hour"class). This class occurs near the end of their program. As you will see, arming them with these habits is not as easy as it may seem.

Background

Elsewhere I describe in detail the influence of two main bodies of research on the design of my mathematics education course (Goodell, 2006). The first concerns teaching for understanding using constructivist pedagogies, and the

J.E. Goodell (✉)
Cleveland State University, Cleveland, OH, USA
e-mail: j.goodell@csuohio.edu

S. Schuck, P. Pereira (eds.), *What Counts in Teaching Mathematics*, Self Study of Teaching and Teacher Education Practices 11, DOI 10.1007/978-94-007-0461-9_8,

second, reflection on practice as a vehicle to promote learning about practice. These two theoretical frameworks underpin all the activities in which I engage the students throughout the entire semester.

In the second class meeting every semester, my students read and discuss at length an article by Carpenter and Lehrer (1999), which proposes five forms of mental activity from which mathematical understanding develops. These are (a) constructing relationships, (b) extending and applying mathematical knowledge, (c) reflecting about experiences, (d) articulating what one knows, and (e) making mathematical knowledge one's own (Carpenter & Lehrer, 1999, p. 20). In groups, my students examine the five forms of mental activity that lead to the development of understanding, and we discuss how to recognize understanding and misunderstanding. A rich problem that necessitates engagement with each type of thinking is completed, and the features of the task that made this activity possible are distilled. Students are also required to post their reading reflections and respond to another student's reflection on the Blackboard website that I maintain for the class. These activities set the stage for discussions about teaching for understanding that follow throughout the semester.

Future class meetings involve critical incident discussions, hands-on problem-solving tasks, lesson presentations with peer evaluation, and a focus on examining each of the NCTM principles, content and process standards (National Council of Teachers of Mathematics (NCTM), 2000).

Methods

Self-study of Teaching

As noted by LaBoskey (2008) self-study research has an established track record as a viable field of educational research. Confirmation of this was also provided when the American Educational Research Association (AERA) established a panel on Research and Teacher Education to review the literature and recommend future research directions for teacher education (Cochran-Smith, 2005). In order to be included in the panel's review, self-studies were treated as interpretive studies "in that they had to have clear research questions, explicit discussions of processes for data collection and analysis, and full descriptions of the conditions and contexts in which they occurred" (Cochran-Smith, 2005, p. 223).

I began my self-study during my first year as an assistant professor in a tenure-track position. Like most new tenure-track faculty, I was concerned about the tenure clock, and wanted to ensure I had something else to publish apart from my recently completed dissertation. Self-study was a natural choice as it required me to reflect deeply on my teaching, something I asked my pre-service teachers to do as well. I felt then, and still do now, that I would not ask my students to do something I would not do myself. If I want my students to develop reflective habits, I should exemplify those habits as well. My growth as a teacher educator is directly connected to my engagement with self-study; similarly, their growth as teachers stems from their own self-studies.

As noted by Schuck (2002), essential to the success of self-study as a pathway to reform is an understanding of the context in which students are working and of their beliefs about mathematics education. My self-study meets both of these requirements. First, I am deeply embedded in the local school districts in which my students are placed for their field experiences. Not only have I been the main university supervisor for both major field experiences, I have also directed numerous grant-funded projects that have provided professional development in reforming mathematics and science teaching and learning for over 600 teachers in these districts since I joined the university in 1999. In addition, my own career began as a high-school mathematics teacher in a high-poverty urban school. Second, I engage my students in deep reflection throughout the semester, and require them to read, discuss and write about their beliefs continuously. These two avenues provide me with a good understanding of the contexts of teaching my students are involved in.

The goal of my self-study has always been to improve my practice as a teacher educator, but I also believed then, as now, that reflection alone was not sufficient. I needed data to reflect upon, and so I used my students' written reflections as one source of data for each of my successive self-study papers (Goodell, 2000, 2002, 2006). But the focus of these papers has evolved over time as a result of engagement in the interactive process of meaningful self-study. In the beginning, I wanted merely to ensure that my class activities were addressing student teachers' concerns about teaching, which they expressed through their choice of critical incidents. Subsequently my research question evolved to focus on what was learned about teaching for understanding through reflecting on critical incidents. My research question in this paper is how do pre-service teachers' beliefs about teaching mathematics, as expressed through their written personal philosophies of teaching, evolve throughout the semester and what factors influence that evolution.

Data Collection and Sources

This study took place in a medium-sized (16,000 students) public university in a large urban mid-western city. I have taught the secondary mathematics methods course at my university for 10 years and have data for 151 of the 158 students whom I have taught over that time period. My students were mostly non-traditional students, many of whom were pursuing post-baccalaureate certificates or master's degree programs. There were 23 students in alternative certification programs (AC). Masters' degree candidates in the Master's of Urban Secondary Teaching (MUST) program totaled 40 students. Post-baccalaureate students (PB), all of whom already had an undergraduate degree, totaled 35 students. Undergraduate (UG) students, who were concurrently earning a degree in mathematics and a teaching license, totaled 49 students. Three students were visiting (VS) from another university and not part of any program. Figure 8.1 below displays the gender breakdown of each group.

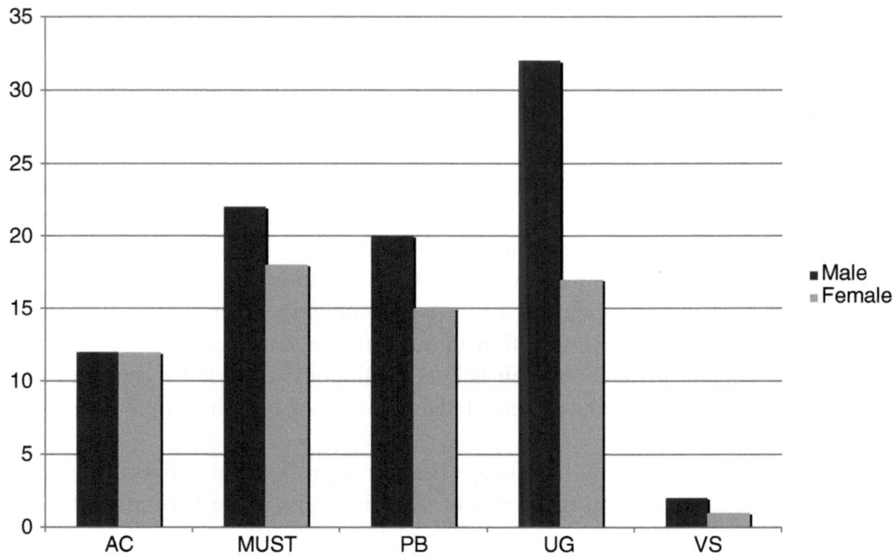

Fig. 8.1 Student characteristics by program and gender

The mathematics content requirements for secondary teaching certification are the same for both undergraduates and graduates; so those graduates whose undergraduate degrees are not strongly mathematics focused had to take a considerable number of mathematics content courses, thus ensuring that their mathematics content background was comprehensive. In the second-to-last semester of the program, students enroll in my course, which pairs a field experience of 180 contact hours, known as "practicum" with a 4-semester-hour mathematics teaching methods. In the semester following this course they have a 15-week full-time field experience, known as "student teaching", which involves much more teaching and usually occurs in a different school to their practicum placement.

The data for this study consist of the written final papers submitted by the students at the end of the semester. Due to the large volume of data contained in the 151 final papers, a random sample of 25% (37 papers) was analyzed. The sample demographics were representative of the population, and included papers from every class I taught over the 10-year period.

The format for this paper was clearly explained in the course syllabus, and did not change significantly over the first 9 years. Below is how it appeared in the syllabus from 1999 through 2008 with minor modifications for clarity.

1. Revise your initial philosophy constructed at the beginning of the semester, indicating what things have grown or evolved, and including specific references to the literature we have read throughout the semester.
2. How did the systematic critical reflections on your teaching through the critical incident discussions and papers influence your growth?

3. What were some of your main concerns, struggles, obstacles, questions, excitements and "ahas"? How did you resolve those obstacles or conflicts in your teaching? What are some of those conflicts that you are still searching to solve?
4. What were some of the other influences on your professional growth this semester?
5. What is your personal philosophy of mathematics teaching and learning now?

In 2009, I changed the final paper to include more specific reflective questions. Questions 1, 2 and 3 remained the same. The remaining questions were as follows:

4. Describe two or three realizations you have made about adolescents this semester.
5. Describe one or two things that you discovered about your own perceived strengths and weaknesses as a teacher.
6. Describe one or two things you learned about classrooms and schools this semester.
7. Describe two or three realizations you have made about teaching this semester, and outline what influenced your growth in these areas.
8. What were some of the other influences on your professional growth this semester?

I graded the final papers using a rubric that was provided in the syllabus at the beginning of the semester. We discuss in class the expectations for this assignment, and I make it clear to my students that I want them to be honest in their answers, as well as to address each of the questions listed. The majority of students score very well on this paper, and those who do not are invited to resubmit a revised version to earn a higher score. This paper contributes 10 or 15% (there are different requirements for undergraduate and graduate students) to their total score for the whole course.

All papers were submitted electronically and entered into the qualitative data analysis (QDA) software, NVivo. Computer-assisted QDA software (CAQDAS) allows the researcher to easily attach codes to text and retrieve all instances of similarly coded text together in one document. Patterns that might apply to one group or another, for example if comparing males and females, can be more easily identified once text coded into one theme is collected together. The software does not do any analysis, the researcher does. The main advantage in using CAQDAS is the ease of managing and organizing data. Since I had 37 papers of approximately 5 pages each, trying to manually manage this amount of data would be very difficult. For more information about the benefits of using CAQDAS, see Silverman and Marvasti (2008).

Six broad codes based on the research questions were used initially to categorize the data, and these also loosely corresponded to the responses students made to each of the questions on the final paper. Not all students responded directly or completely to each question, which made it impossible to make categories that were exactly linked to each question on the final paper.

The six main categories identified in the data are discussed in detail in the next section of this paper. These are changes in teaching philosophy; realizations and "ahas"; influences of the critical incident reflections; other influences on growth; main concerns, struggles, obstacles and questions; and changes in teaching.

Results

What Changes Are There in Pre-service Teachers' Teaching Philosophies?

Of the 37 student papers analyzed, 25 reported that their personal philosophy of teaching was still the same, with some minor changes and additions mostly attributable to gaining practical teaching experience. For example, one student writes:

> My personal philosophy was well written and full of great ideas and hopes for my teaching profession. I still stand by everything I stated in my initial philosophy but I feel that after this experience that I would have to change a few things around and add a couple more important ideas I have developed during my practicum experience. (Bianca, Female UG)

To have such a large proportion (67%) of the analyzed sample stating their philosophy was basically unchanged was surprising to me given the structure and tone of the discussion in class, which was almost always in support of a constructivist philosophy. However, on examining the initial philosophy statements and the restatement of the philosophy in the final papers, most students had initially expressed rather idealistic views. The reality of real and sustained classroom teaching practice impacted their ideas about what is actually possible, so their philosophy was not so much transformed or changed but reinforced in various ways. A good example of this sentiment is expressed below:

> I do not intend to revise my philosophy, which is still in its original form above. ... I still believe in what I wrote a few months ago. Reality has taught me that some things are simply impossible due to time constraints but I still strive to achieve them as much as practical. I do not give up easily and hope that I do not burn out in this profession as many teachers have. (Daniel, Male PB)

Only four students (10%) stated that their philosophy of teaching was significantly changed over the semester. One student stated:

> The practicum experience and this course have dramatically shaped my teaching style and provided a wealth of strategies, tactics, and material that I will use throughout my career. (Barry, Male UG)

Areas of Growth: "ahas" and Realizations

Despite the majority of students stating that their beliefs had not changed significantly, the analysis generated a very long list of items my students had learned or realized but were unaware of prior to the practicum experience. These realizations focused on seven areas: teaching mathematics, teaching in general, classroom management, students, curriculum, policies and school contexts.

Almost half of their comments (15 out of 33) were in the teaching mathematics category. Understanding the importance of teaching for understanding was one subcategory, illustrated by this comment.

> The most important thing I have learned this semester is that we should teach for "understanding" (Carpenter, 1999). In my experience, I can tell when a student "understands" a concept because they can apply it, summarize it, analyze it, and extend it. In other words, a student that "understands" a topic has learned it and will be able to use it later. This type of learning negates the problems of inadequate prerequisite knowledge because when you truly "understand" a concept, you never forget it. (Calvin, Male UG)

Understanding that many students do not find mathematics as easy as they did is another important area of growth, illustrated by this comment.

> Growing up I excelled in math and so did my peers. I was in honours classes, so I never really got to see how many people did not like math and struggle with the subject. This is my second realization; many people do not like mathematics so in order to teach it effectively, the teacher must relate the material to the students. (Carlos, Male UG)

Another area of growth concerned putting inquiry-learning in practice with their own students.

> The *main* "*aha*" was, what is really meant by the constructivist approach? Even though I have tried different techniques in the classroom I did not understand the significance of letting the student come to the conclusion or create the formula for themselves. I thought you designed a task with the goal or formula clearly stated and let them prove it, really that is no different than asking the student to do a problem. Now I see the benefit of finding the method along with the formula. It is the discovery process that makes the activity meaningful and gives the student the confidence to try more experimental methods of creating patterns, deriving formulas and solving problems. (Denise, Female CORE)

What Influenced These Changes

Critical Incident Reflections

Of the 37 papers analyzed, 33 made substantive comments about the critical incident reflections and discussions, and all of these comments were positive. Most were about the exposure to many different situations that these discussions provided them, or about how the ideas they gained from this would be useful to them in their future teaching careers, as indicated in the following comments.

> The critical incidents that we discussed in every class showed me that I was not alone out there, which is what I had felt for a good part up to that point. Others had encountered situations similar to mine; they had the same problems, and on occasion had tasted the same wine from the cup of success that I had tasted. . . . The discussions we had, whether in person or through the postings on WebCT, demonstrated a wide range of viewpoints about different issues in mathematics education, and I like to think that I was able to gain a little bit from each person and each discussion. Most importantly, I hope to use as many of those ideas in my future career as possible, which will make me the better for the experience, and give my students a better chance to succeed in the school and in the game of life. (Brad, Male UG)

Other comments were about the benefits of reflective practice as a vehicle to improving their own teaching, as expressed below.

> I thought the critical incident discussions were crucial in making me slow down and really think about what happened that day, but more importantly I was always thinking of how I could do it better next time. (Dora, Female UG)

> On several occasions, I was able to use my reflections as a tool to discover why my students did not completely understand concepts that I had presented to them. In these cases, I was able to represent the material after reflecting and use different techniques to make sure the students understood the concept. (Dale, Male UG)

However, only a few responses were truly meta-cognitive. Take, for example, the following comment from a student who realized that the process of reflection changed his interpretation about what happened.

> The pace of a classroom moves much quicker than a teacher is able to reflect on in the classroom. As a result, I found that writing about my experiences, either in my Critical Incidents reports or on the discussion line on WebCT, was one of my most powerful tools to really learn about the true ramifications of my experiences. Oddly enough, I would choose an experience to write about, primarily because it was intense, and believing that, I knew what I would write about it. Through the process of reflection, most of my incidents turned out to teach me something entirely different than what I originally thought. Now, I understand why those teachers who make reflections as a constant practice of their profession become the best at it. (Adam, Male PB)

Clearly, having to reflect on his classroom experiences over the 15 weeks of practicum helped Adam to think deeply about his thinking, and appreciate the true value of reflection. This level of meta-cognition was only evident in two of the 37 papers analyzed.

Other Influences

Responses in this category included many activities done in my class such as assigned readings, hands-on investigations, presentations and reflective writing. Also included were the mentor teacher, other teachers they encountered during practicum, their university supervisors and their family. There were 53 comments in this category with 11 comments about class activities, 9 about readings, 11 about their mentor teacher and 5 about other teachers at the school.

Comments about class activities mostly focused on the group work and hands-on discovery-type activities we do. Most of my students have experienced very traditional expository teaching styles throughout their mathematics education, culminating with their most recent experiences at the university level. Discussions in class clearly indicated that even for these successful mathematics students, their experiences did not promote deep understanding of mathematical concepts, therefore a significant amount of time in my methods class is devoted to involving them in doing and discovering mathematics. The following comment captures this sentiment.

> So many times in my own education I received traditional instruction that even today I have difficulty connecting concepts to procedures. Too many times, I take the prescriptive approach to problem solving. Witnessing this contrast in approaches within the dialogic community and within our methods class, really made me appreciate the importance of constructing knowledge and making connections more than ever. (Bridget, Female MUST)

Class readings were another major sub-category students mentioned as influencing their growth. As noted earlier in this paper, not only did the students have to read the assigned reading before class, but after discussing it in class they were required to post their reflection on the web-based course management tool I used (either Blackboard or WebCT) as well as commenting on another student's posting.

> I responded to article number four and I felt that article was an excellent read because it deals with something teachers either look past or forget, the fact the students do not know why or what they are learning. I shared all my views from this article and then I got to go and read everyone else's views on my response. That aspect was nice because in case I miss a valid point or something did not occur to me one of my colleagues could have brought it to my attention. (Bianca, Female UG)

Finally, the mentor (the teacher in whose class they were completing their field experience) and other teachers at the school were important influences on their growth, mostly positive, but some negative. For example:

> The Math Department has been great, especially my mentor and the head of the department. They're always there if I have a question, regardless of what it's about, and they're always willing to supply some tips, advice, or wisdom to help me with any phase of the educational process. They have also included me in discussions on such topics as: four day work weeks, budget cuts, behaviour issues and procedures, and the possibility of changing over to small schools. They definitely made me feel like I was a member of the faculty and welcome at this school. (Brock, Male UG)

However, it was not always the case that the mentor teacher was a good role model. Some role models provided examples of what not to do.

> First, it was imperative for me to watch my mentor teacher teach. Where she was more rigid in her ways and was more traditional in her methods, I tried to be a little more creative and innovative. (Dora, Female UG)

The availability of excellent mentors has been an ongoing problem for my program, one that is not easily solved.

Main Concerns

The concerns that my students still had at the end of the semester provide insights for me as a teacher educator on issues that need to be addressed in our program in order to better prepare our graduates, bearing in mind that they still have one more 15-week full-time field experience in the following semester.

The concerns fell into the same sub-categories as the realizations except for the addition of a category for assessment, and no comments in the Policy category. Table 8.1 shows a summary of the number of comments in each subcategory.

These findings are in accord with what would be expected from pre-service teachers in their first extended practicum. Concerns about their general teaching skills included such categories as motivating students, lesson planning, making connections with students, time management and overcoming nerves. Motivating disinterested students is a perennial problem and is often a cause of teacher burnout, particularly in high poverty urban settings where most of my students are placed for practicum. Betty's comment below shows that she is well aware of the dangers of trying to "save" every student, a conversation we often have in class during the critical incident discussions that take place every class meeting.

> One conflict that I am still searching to solve is how can I interest uninterested students? The other is how can I help kids care about math? I think experience will help lead me toward resolving these questions. Unfortunately, I think the answer is some students will never be interested and some students will never care. This conflict will never stop bothering me, and I will continue to work on the resolution throughout my teaching career. However, I will not let this conflict deter me from teaching as best I can to students and reaching as many students as I possible can. (Betty, Female PB).

Classroom management skills will always be a big concern for pre-service teachers. This sub-category included comments about creating a positive classroom environment, managing group work and finding the right balance of discipline. As noted by one of my students:

> Off the bat I found that classroom management was going to be one of my biggest problems. The students saw me more as a substitute teacher and tried to test me. Since I was only there for a half of a day I was unable to hold kids after school to talk with them about their behaviour, so I could only do things in class while continuing to teach the rest of the class. (Burton, Male UG)

Table 8.1 Concerns Remaining

Sub category	Number of students	Number of comments
Teaching mathematics	11	11
Teaching	23	25
Assessment	7	9
Management	17	17
Contexts	10	11
Students	5	5
Curriculum	4	4

Concerns about teaching mathematics focused on how to implement constructivist learning environments, adequacy of content knowledge and familiarity with classroom technology. One student commented on her frustration in implementing a constructivist learning environment.

> I did have a tough time this semester trying to live up to my own philosophy. I think I expected too much too soon by wanting every lesson to be taught in a constructivist manner, along with activities to stimulate interest and help students create their own understanding. Needless to say it was an impossible goal and I often felt frustrated. (Fiona, Female UG)

Changes in Teaching

This category contained only seven comments, and all were concerned with technical aspects of teaching such as adjust teaching to students' learning styles, better lesson preparation, calling on students, challenging students, increasing think time and asking more questions. The lack of comments in this category may point to the fact that the majority of my students do not have well-developed teaching skills overall, and are therefore unable to determine what needs changing when they know so little about the impact of particular teaching practices on student learning.

Summary

In this section, I have presented a selection of quotes that illustrate areas of growth for my students, the influences on that growth, and the concerns that remain at the end of the semester. Seven major findings emerged:

1. Their personal philosophies of teaching were remarkably stable over the duration of the course.
2. Becoming an inquiry teacher must be preceded by learning through inquiry, and both of these are time consuming and difficult processes.
3. Reflection on critical incidents allows the frustrations and joys of teaching to be shared, and can help teachers feel less isolated.
4. There were very few meta-cognitive responses, despite my best efforts to encourage this.
5. Class activities were generally perceived as helpful, especially to provide an image of inquiry learning in practice.
6. Excellent mentors are a crucial part of the process of preparing new teachers. Identifying and educating them is critical to the success of reforming mathematics education.
7. Technical aspects of teaching are far more important to beginning teachers than philosophical considerations.

In the next section, I will discuss what this tells me about my teaching and the changes that have resulted from that learning.

Discussion

The implications of these findings for my teaching are many. Confronted with the large percentage of students who stated that their philosophy remained essentially unchanged over the semester was surprising to me given the nature and tone of classroom discussions, which often seemed to indicate their acceptance of constructivism as the theory on which mathematics learning should be based. Further reflection on this has forced me to accept that weekly classroom discussions over a semester are not sufficient to impact beliefs or philosophies, especially if traditional views of teaching and learning mathematics are held by mentors and other teachers encountered in the practicum placements. This implies that finding suitable placements is something I need to think more deeply about and spend more time on. It may also mean that students tell me what they think I want to hear in class, but are more honest in their writing.

The second important realization I have made is that the reflective thinking (and writing) is much harder than it may seem. For it to be truly transformative requires you to be brutally honest with yourself, a task that many people are not accustomed to or willing to do. The critical incident discussions and reports are very helpful to the students in many ways, but very few of them see "what to me is the true value", which is that developing the habit of thinking about teaching changes the way you teach. Only two of the 37 papers analyzed showed this type of meta-cognition. My previous self-study highlighted this fact, and from that time on I tried to improve significantly the feedback I gave students in class discussions and on written work to help them think about the implications of the incident for their teaching (what they learned). I now plan to do additional analysis of the critical incident reports, comparing the more recent reports to the earlier reports for any difference in the quality of reflection. I will also make the focus of my next self-study to investigate what more I can do to improve my students' meta-cognitive abilities through reflective discussions and writing.

One aspect of my teaching that is reinforced through my self-study is the importance of providing multiple opportunities in class to experience inquiry learning. However, I have also realized that teacher educators should be cautious about making inquiry-based activities the sole focus of a teaching methods course. Clearly it is impossible to provide experiences that encompass the whole high-school curriculum. However, if the beginning teacher is able to think critically about his or her teaching, it is not necessary to "cover" everything.

Reading over and analyzing the final papers my students have written over the past 10 years has reminded me that learning to teach is a very complex task, and that constructivist learning principles apply equally well to learning to teach mathematics as to learning mathematics. Learning to teach must include significant ongoing teaching experiences that are supported by structured opportunities to reflect on and learn from these experiences – essentially to construct knowledge about teaching. However, although I read (and graded) all of these papers at the end of each course, and understood their concerns, it was not until I wrote this chapter that the breadth of issues my students wrote about became clear. Many of their concerns are things that will improve significantly with experience (such as classroom management), and that will benefit from having a longer time period to reflect. It is only natural that with so little experience to draw on, most beginning teachers are focused on the technical aspects and skills of teaching that all experienced teachers know take a number of years to perfect. The more experienced a teacher becomes, the less he or she recalls about those first few difficult years, making it harder to identify with the struggles beginners face. Ensuring mentor teachers understand this should definitely be part of the mentor selection and training process.

Similarly, it was not until I examined the final papers across the years for evidence of meta-cognition of reflective practice that I recognized how rare it was. It would seem that I am over confident that reflecting on teaching will automatically continue past my course given the lack of evidence that they really got the point of it. I see this as analogous to ensuring that at the end of an inquiry-based mathematics learning activity, students understand the concept being investigated and that formal mathematics is connected to the manipulative, tool or situation under exploration. Therefore, I will continue my self study toward ensuring that all of my students understand that the ultimate goal of critical incident reflections and discussions is not just to learn from each other's experiences, but to develop the habits of mind that make reflective practice an essential part of their daily routine. Perhaps assigning this chapter as a reading early in the course will help make my goals explicit from the start and better set the stage for my students to gain a deeper appreciation of the reflective writing and discussions they engage in over the course of the semester.

Conclusion

Over the course of my higher education teaching career, I have taught mathematics methods courses to pre-service teachers in three universities and two countries, each with significantly different requirements for graduation and licensing. In addition, I have been involved with in-service teacher professional development for most of that period. The one thing that almost every evaluation of any of these activities highlights is the great appreciation of and need for teachers to discuss their teaching practices and learn from each other. In both countries, teaching high school mathematics is essentially a solitary activity, in that there is little opportunity and often no desire, to share with colleagues in the workplace, often because of time constraints but also because of the culture of secondary schools. I have often remarked to my pre-service teachers that this class (their only true mathematics education methods class in the current program) is probably the only time they will ever have the luxury of sharing critical incident discussions with a large group of colleagues. They have often remarked to me (and shared on course evaluations) that this class is the most valuable one they have had throughout their program, and that there should to be more like it, earlier in their program, and extending throughout the program, not just at the end.

In 2004 I learned about a teacher preparation program that embodied many of these characteristics. I became convinced that this is what I wanted to implement in my own university. There were significant differences between this program and our existing program, so at first glance it seemed like the impossible dream. However, approximately 3 years later, a call for proposals to replicate this program was issued by a national non-profit organization set up for the purpose of improving the quantity and quality of mathematics and science teachers across the USA. I convinced my colleagues to apply for this funding, but we were unsuccessful in our first attempt. Two years later, we applied to the National Science Foundation and the US Department of Education, and we now have approximately $5,000,000 from these funding agencies, the original non-profit sponsor and committed internal funds to implement the program. This past year has been spent planning and preparing to start the program in August 2010, and as noted earlier, it is a radical departure from our existing program.

There are many features of this new program that have the potential to improve the preparation of our graduates and help them to become reflective practitioners. All of the education courses are set in the context of teaching mathematics and science, and include a focus on constructivist learning theories and teaching methods, inquiry lesson planning and implementation, research methods and data analysis, history and philosophy of mathematics and science, and project-based instruction. Field experiences commence in the first course in the program and continue throughout the program. Much greater induction support will be provided, and this will enable me to ensure that the critical incident reflections are continued past the end of the program. This longitudinal and contextualized focus addresses the concern my students have often raised that they only get to engage deeply in thinking about teaching mathematics in the final year of their program. In addition, as the director of the new program, I will have input into the content and structure of all of the coursework, and I will ensure that reflective writing and discussion about practice will be a focus throughout the program. I am also hopeful that the course entitled "Perspectives on Science and Mathematics" will significantly impact my students' teaching philosophies. As I noted earlier in this chapter, one finding of this study is that my students' beliefs about teaching mathematics are remarkably stable over the course of the semester. One possible reason could be that they believe mathematics to be a fixed body of knowledge that can be transmitted to students. Studying the history and philosophy of the development of mathematics will help them more fully appreciate that mathematics is a human endeavor and is best learned through investigation not transmission.

Engaging in this longitudinal self-study research has enabled me to grow professionally in a way that would not have happened without it. Like other mathematics teacher educators (Schuck, 2002), I have not had a critical friend with whom to discuss and analyze my data; however the process of writing each of my self-study papers has been my critical friend. With the introduction of our new program, it will be necessary for me to change the focus of my self-study to encompass the whole program, in essence program evaluation. I will also have numerous colleagues to collaborate with as critical friends.

When I first started using critical incident reports in my methods classes, I did not make my goals clear to my students, mostly because I had not clarified them for myself. Self-study has provided the method and motivation for me to closely examine and refine the goals I have for the students in my course and ultimately for the program as a whole. Being explicit about your goals and designing courses and experiences to enable your students to achieve those goals is much harder than it seems unless you are constantly evaluating and modifying both simultaneously. I am looking forward to continuing this process with my new courses and program.

References

Carpenter, T. P., & Lehrer, R. (1999). Teaching and learning mathematics with understanding. In E. Fennema & T. Romberg (Eds.), *Mathematics classrooms that promote understanding* (pp. 19–32). Mahwah, NJ: Lawrence Erlbaum Associates.

Cochran-Smith, M. (2005). Teacher educators as researchers: Multiple perspectives. *Teaching and Teacher Education*, *21*, 219–225.

Goodell, J. E. (2000). Learning to teach mathematics for understanding: The role of reflection. *Mathematics Teacher Education and Development*, *2*, 48–61.

Goodell, J. E. (2002). *Learning to teach for understanding through reflecting on critical incidents in an integrated mathematics methods and practicum course*. Paper presented at the annual meeting of the Association of Mathematics Teacher Educators. San Antonio, TX.

Goodell, J. E. (2006). Using critical incident reflections: A self-study as a mathematics teacher educator. *Journal of Mathematics Teacher Education*, *9*(3), 221–248.

LaBoskey, V. (2008). The fragile strengths of self-study: Making bold claims and clear connections. In P. Aubusson & S. Schuck (Eds.), *Teacher learning and development: The mirror maze* (pp. 251–262). Dordrecht, The Netherlands: Springer.

National Council of Teachers of Mathematics (NCTM) (2000). *Principles and standards for school mathematics*. Reston, VA: National Council of Teachers of Mathematics.

Schuck, S. (2002). Using self-study to challenge my teaching practice in mathematics education. *Reflective Practice*, *3*(3), 327–337.

Silverman, D., & Marvasti, A. (2008). *Doing qualitative research: A comprehensive guide*. Los Angeles, CA: Sage Publications.

Chapter 9
Opportunities for Learning – A Self-Study of Teaching Statistics in a Mathematics Learning Centre

Sue Gordon

Introduction

The context of my self-study is my work for a Mathematics Learning Centre, a small department in a major metropolitan university in Australia. I am one of three mathematicians who make up the continuing (permanent) academic staff and who are committed to helping students understand and appreciate the mathematics they are studying. The Mathematics Learning Centre was established in 1984 with the aims of increasing access to mathematics and improving the completion rates for students studying mathematics and statistics at university. The academic staff in the Centre develop and teach bridging courses, develop resource material and provide ongoing support for eligible students of the university – those who are deemed at risk in their introductory mathematics and statistics subjects. We assist students in a wide range of courses by means of individual tuition and weekly small group workshops, which supplement their mainstream tutorials. Typically, we offer help in mathematics and statistics to over 600 students annually (Nicholas, 2010). Students attend the Centre voluntarily, sometimes for many hours each week.

Many of the students I teach lack the prerequisite background in mathematics. Some may be termed reluctant learners as they do not have an intrinsic interest in mathematics. Psychology students who attend the Centre for help with statistics are of particular interest to me. One of my major concerns is to develop teaching methods that are innovative and appropriate for these students – arguably among the most anxious and unappreciative of university students concerning the study of a mathematical subject. One such student, who attended the Centre regularly, wrote this summary of her feelings about learning statistics:

> I don't feel confident with statistics.
> I don't plan a career that would involve statistics.
> I don't enjoy statistics.

S. Gordon (✉)
The University of Sydney, Sydney, Australia
e-mail: sue.gordon@sydney.edu.au

S. Schuck, P. Pereira (eds.), *What Counts in Teaching Mathematics*, Self Study of Teaching and Teacher Education Practices 11, DOI 10.1007/978-94-007-0461-9_9,

These students seem to me like travellers who have been diverted from their route to this strange "Land of Statistics". Some of them do not know what they are doing there. They do not speak the language, and they follow desperately and blindly any sign which may get them out. ("Do you ALWAYS do t-tests in statistics problems?" asked one student.) My job as a teacher is to provide a map that enables students to choose ways to their destinations and to enjoy the journey. Self-study helps me draw this map.

Self-study methodology is an evolving methodology and self-study practitioners are involved in a "continuous and communal conversation" (Samaras & Freese, 2009, p. 11) about the defining characteristics of the philosophy and methods underpinning self-studies. Loughran (2006) proposes that there are three (unordered) levels of self-study: personal, collaborative and institutional. My self-study fits into the "personal" category with a focus on the self in "trying to better understand practice" including dilemmas, tensions and issues and to explore ways of reconceptualising practice as well as "offer access" to the manner in which knowledge of practice has developed (p. 50). That is, I try to make transparent the insights afforded by aspects of my practice and how I translate these insights into effect. The self-study literature has alerted me to two ideas in particular, which seem to be critical and resonate with my thinking: the idea of a purposeful questioning of one's own assumptions about teaching and learning and the ongoing reframing of practice as new understandings emerge. My self-study is not complete and closed but continually unfolds with reflexive practice. Moreover, this unfolding is not a smooth course of insight and progress.

In this self-study I reflect on my experiences and aspirations while teaching in the Mathematics Learning Centre. I aim to illuminate ways in which selected activities and dimensions of my work have provided opportunities for me to develop my practice. I outline how early research with students at the Mathematics Learning Centre provided transformative insights and started me on an ongoing spiral of teaching, research and scholarship. I highlight findings that have been pivotal to my understanding of teaching and learning or have been a source of motivation and inspiration. I introduce the idea of incongruity as a potentially useful concept for self-study.

Taking account of the perspectives of others to develop one's own understandings and reframe interpretations of experience is a long-standing principle of self-study methodology (Loughran and Northfield, 1998). Often the "other" is a critical friend (Schuck & Russell, 2005). My insights derive from paying attention to the views and observations of: the students whom I teach; co-researchers; and participants in research projects on teaching and learning in higher education. Scholarship has enabled me to submit interpretations and ideas from research projects to the scrutiny of others and accordingly advance my own thinking about the findings. However, the stories of my own growth as an educator have been in the background, unexplored and unquestioned. Bringing these stories to the foreground is my aim in this chapter.

Early Research as an Impetus to Reframing Practice

Theoretical Lens

My doctoral study (Gordon, 1998) of the theories of Vygotsky and Leont'ev (for example, Leont'ev, 1981; Vygotsky, 1962, 1978) was an early stimulus for exploring student learning. Activity Theory posits a systemic view of human behaviour in which an individual's goals and subjective perceptions are interwoven with socio-historical factors. Activity Theory views human learning as active behaviour mediated by interactions with the social world. We learn by our engagements with other people and with our cultural and historical environments.

My study of the complex, and initially alien, Activity Theory led me to an awareness that the ways students learn mathematics or statistics are inseparable from an intricate web of personal, social and cultural factors. This is important because students' activities determine the quality of their mathematical learning and emerging knowledge. The way a student learns statistics reflects her personal story, a history of her own experiences, which have left their mark. In addition, each student's learning reflects values and beliefs of communities and cultures that are significant to her. These contribute to her ways of understanding statistics and of interpreting the context in which she encounters it. According to Vygotsky (1981, p. 162): "We could therefore say that it is through others that we develop into ourselves and that this is true not only with regard to the individual but with regard to the history of every function".

A student's understanding of statistics is refracted through the prism of community knowledge and the mechanism for this is social interaction. The links or bridges between individual abilities and the broader society or culture are cultural tools. Technology and, most of all, language are examples of cultural tools. The experience or knowledge of the past generations is passed on through social interactions and cultural tools and is renewed through these. Through mastering cultural tools we develop our minds (Vygotsky, 1978).

Investigating Purposeful Questions

As I immersed myself deeper in theories of Vygotsky (1962, 1978) and Leont'ev (1981), I tried to apply their ideas to reframe my approach to teaching. Drawing the map for students became a lot harder. My focus had been on how to enable students to appreciate statistical concepts and how to introduce students to the discipline of statistics and its applications in psychology. Hence, my primary focus was on the nature of the information. My teaching approach was empathetic, listening to students, encouraging them and showing them how they were progressing.

Activity Theory alerted me to thinking about the student acting in her social and cultural world, that is to thinking about what the learner does; why she takes those actions; and how her actions relate to the learning arena. I realised that while I, as a teacher of statistics, may try to emphasise the usefulness of the subject, no one can persuade a student of the power of a tool. A student's interpretation of the use of statistics depends on the student's experience of its functionality in terms of what learning statistics means to the student. In particular, the way a student monitors and controls her ongoing cognitive activity depends on what she is trying to achieve and what she deems to be the criteria of success.

My reflections generated questions as to what the meaning of learning statistics could be for students.

- Firstly why is the student learning statistics? What is it that energises the learning activity? What are the goals?
- What are the student's experiences of statistics?
- What is the social and institutional context in which the learning occurs? And how do the student's perceptions of this context affect his or her approaches to learning?
- How does the student evaluate and reflect on the task?

To try to answer these questions, I began by investigating the learning of psychology students who sought assistance with statistics at the Mathematics Learning Centre. The excerpts quoted in this section are from interviews and written surveys for an exploratory study described in my doctoral thesis (Gordon, 1998) and earlier published articles related to this research. All names used are pseudonyms. I was not involved with the students' formal assessment in any way but worked extensively with them during the year they studied statistics, and was able to develop a close personal relationship with these students. As will be demonstrated below, attempting to answer these questions generated insights but also new questions.

Goals

Leont'ev (1981, p. 60) proposes that goals are the "energising function" of the student's activity. Yet often, in practice, if not in educational theory, the goals of the students are assumed to be irrelevant to their actions and simply ignored. The examples below illustrate that students produce differing goals in the process of participating in the activities and that these goals are integral to their approaches.

Vicki expressed the idea that statistics, far from being essential to the application of Psychology, is an unwelcome intruder. In a survey she wrote:

> I don't even see the point. In psych why must maths infiltrate itself? Studies have shown that those who have high maths abilities have low or poor communication & perception skills – shouldn't psychologists be exceptionally perceptive & able to communicate well? It seems that if there aren't silly numbers to justify things then they aren't plausible in our computer/maths/science promotive society.

In contrast Tessa reported an understanding of the use and application of statistics:

> It's very interesting – how statistics moulds itself into Psychology. Reading through the texts in Psychology there's a lot of statistical data; data that's been ordered in a statistical way to make it more of a hard science. In (school) maths there wasn't that much application to things in real life. A lot of it was just formula based, following the formulas, plugging in the numbers. With statistics, however, you've got an aim. In a real life situation, in society, statistics is like a tool to analyse whatever happens when you do experiments.

These two students report contrasting goals that indicate quite different approaches to learning statistics. For Vicki outright resentment at having to study statistics would seem to prevent the task from assuming any meaning. Tessa, however, reports an awareness of the meaning of the knowledge. Their contrasting perceptions resonated with my awareness about the different positions, emotions and attitudes of students who came to the Mathematics Learning Centre for help in learning statistics.

A third student, Norman, expressed a positive view of mathematics, whether in the everyday context or more theoretical.

> There is maths which is simple numbers perhaps and maths which is concepts. Now I like maths which is numbers, I always add up numbers in my head in supermarkets, things like that. I like to play with numbers.

When pressed for his definition of mathematics he replied:

> It's a discipline or science concerned with numbers and quantification, perhaps. I do philosophy and I'm aware that maths is arguably being used to explain everything in the universe. I like the subject on that level – it's interesting, rational, and abstract.

Hence, Norman studied statistics without anxiety or trepidation. He was comfortable with mathematics at any level, whether concrete or abstract, and viewed it in terms of a game to extend the mind. However his stated goal for studying statistics was simply to fulfil the assessment requirements as reported below.

> I have a very pragmatic approach to university; I give them what they want. Arguably if I could guarantee enough knowledge to get ten (full marks) in the tutorial test and the exam and know that I forgot it all completely afterwards, I'd almost go for that course, "cause that's what they want".

Norman's clear-cut and, perhaps to others, unexceptional account of his aims was an epiphany to me. Suddenly my pleasing but myopic view of the academic landscape and students' activities within the Mathematics Learning Centre changed. Norman had shown me that a highly capable and mathematically confident student was not interested in making his way through the complex terrain of statistical knowledge, as I had taken for granted in my teaching. He was looking for and had found a short cut. Further, this was possibly a rational strategy that took account of the course environment where assessment was often through a set of multiple-choice questions under constraining time limits.

Experiences

The individual's experience of statistics – at school or in other settings – is a critical factor in determining what they think statistics is. University students' expectations of learning statistics in the educational context are often based on their appraisals of school mathematics. These are often regarded very differently to their experiences of "life" statistics. Consider the following extracts from interviews with mature students who returned to studying after many years away from the education system.

Ernest:

> As a hobby for years, I have followed the races. I don't bet very much, but when I do it I am a mathematical punter, I look at the percentages. So I relate to probability. Probability is effectively something that I use. I like the idea of comparing my percentages with those that the bookmaker has on the board, and at that level I have always dealt with statistics.

While no reluctance to engage with "real life" statistics was reported, Ernest indicated a quite different attitude to school mathematics. He said:

> Arithmetic I didn't mind at all, but the moment it became more abstract, the moment symbols entered the scene, without a teacher who could relate it to the practical usefulness of it – he lost me. It became an exercise that I couldn't see mattered much to me, had relevance to what I wanted to know. I would stare out of the window. It went very much past me.

Hettie, a student who had little formal schooling in mathematics, said:

> I had accounting skills from running a business, so I was not completely innumerate.... In second year Psychology, from day one, when we got the (statistics) handout, I was panicked – by the algebraic equations, everything. This was what scientists, astrophysicists do, not what I could do. When he put all that stuff on the board, I just froze. I was like the rat I'd been observing in the laboratory earlier on. I just curled up in my seat and froze in terror.

Hettie's fear was a boulder in her "path" to learning statistics, and one she herself had put there. This alerted me to recognising that affective factors could be as important as cognitive factors in learning statistics and should not be relegated to the background.

Impact of Institutional and Social Context

The Activity perspective on learning statistics highlights that students' approaches to learning statistics are related to contextual factors. For example, where but in a university, would you get students sitting writing down things whether or not they have the slightest idea of what it is they are copying, as Alice related?

Alice:

> In the lectures I was sitting watching him go through the overheads of our notes and if he said: "This is very important". I would underline it and if he said: "Don't worry about these pages", then I wouldn't worry about those pages. I assumed the man knew what he was doing because he wrote the notes!

Alice wanted to understand and use the concepts. However, her difficulties with the mathematics and her perceptions of time constraints and assessment demands strongly inhibited her actions.

She said:

> I thought if I could grasp the basic concepts I would be able to apply it to my exams – but I couldn't. There was something missing and I think it was my comprehension rather than their teaching.... It's in a different language. I don't have a problem with languages – it must be the figures, Sue, I can only assume it's figures. I think the whole subject is really difficult! I just know I haven't understood it.

However, when working with her colleagues in a more relaxed environment, Alice reported a more positive approach.

> I was working with Norman and Sandra – every Sunday night, almost the entire year. We went back over the previous week's lectures and tutorials, and went over it again and again and again, with the help of Sandra's husband who understands statistics. I can do all those things with the others around a table, we can get them right.

Norman reported adopting a surface approach (Marton, 1988) to learning statistics yet he achieved a high degree of success in the assessments:

> I just wanted to pass the course. Practice and repeat is the key. You have to have the basic understanding of when to apply the test, and that is almost formula. They are not looking to see if you critically appreciate because the test questions are very similar to the tutorial questions. Compare the test question to the tutorial question; if they're the same, apply the same formula.

Norman found the multiple-choice examinations harder than the class tests because of the speed with which each question had to be answered. For these, too, he relied on superficial clues, rather than understanding. He said:

> To get away with doing it in one and a half minutes a question – anything less than knowing it really, really well – you were in trouble. I had to guess. A lot of multiple-choice questions are really tricks. If two of them are similar, then usually it's one of them.

Despite Norman's perceptions of the superficial quality of learning required in the context of academic learning, the social context provided by his working regularly with Sandra and Alice resulted in a deeper approach. He said:

> I invariably picked up some understanding anyway. Compared to Sandra and Alice. When I worked with the others as a group then I would try to do more of the understanding.

In Norman's case the different approaches he adopted to learn statistics and the quality of the ensuing knowledge, as he reported it, were interwoven with the social settings surrounding his actions. Together, Norman, Sandra (and her husband) and Alice created an interactive learning environment very different from the lecture-tutorial settings in which they mostly studied statistics and in this environment their activities were directed at attaining a conceptual understanding of statistics.

These reports express the importance of creating a learning environment where students take responsibility for their own learning. For me, this suggests that the first challenge is to let go of the view that I am the sole owner of statistical knowledge and the overriding expert.

Students' Self-Evaluations

This brings us to the most important factor in students regulating their learning: the students' own evaluations of their learning. Now for all students the examination mark is obviously important. However not all students evaluate themselves entirely by this quantitative level as the brief excerpt below indicates.

Sandra:

> It's almost like two separate things in the statistics course we've just done. You could have actually just got the steps and maybe not understood why you were doing it. I wanted to understand what I was doing.

Sandra's evaluation of the subject was in terms of the professional approach that she had achieved.

> While a lot of course material is not relevant for counselling a lot of the attitudes that you learn are important. A sense of professionalism in your approach, not a knee-jerk reaction to things, but sitting back and assessing it. The course teaches you to take many theories in and assess the different theories. And sit back a bit more. People who haven't had that training... they have more... a knee-jerk reaction.
>
> By the end of the year I thought, it doesn't really matter how I go in this exam, I'm not going to let the exam mark dictate to me my knowledge.

Sandra's insightful comments inadvertently highlighted a quandary. Sandra wanted to understand the concepts and assumptions underpinning statistical processes. However, perhaps for some students just "getting the steps" is the only achievable endpoint. I continue to wrestle with this dilemma: how to orient students toward high quality learning without losing sight of the immediate priority, which is to meet the demands of assessment. This is difficult because my work is in a support centre where I have no input into curriculum or assessment. Does my focus on helping all students see the wood rather than the trees – an insistence on students understanding "why" before "how" – disadvantage some students in assessments?

New Understandings and New Questions

This early research is central to my conception of my own practice. It alerted me to the diversity of students' experiences of learning statistics. Hettie's metaphor comparing herself to a laboratory rat demonstrated the strength of her perception of helplessness while Sandra's eventual rejection of the examination mark as a measure of her knowledge was self-affirming. The findings forced me to re-examine beliefs about my role as a teacher of statistics. An unquestioned assumption underpinning my pedagogy was that psychology students *needed* to understand statistics as a basis of the empirical method for their discipline and that my role was to empower the students who came to the Centre to learn to think statistically. Norman's expression refuted this need for himself. I began to probe more deeply into my own reaction to his statements. Was my rejection or even denial of some students' instrumental or short-term goals based on my own need to transform students' outlooks rather than on pedagogical principles? The question: "Whose needs does this endeavour address?" challenges my unthinking complacency.

This early qualitative exploration of students' "activities" (Leont'ev, 1981) at the Mathematics Learning Centre – their needs and motivations, goals and actions shaped within the wider university arena – also led to further questions and started me on an ongoing and extensive program of research. Initially, I expanded my research with students at the Centre to a wider, multi-variable investigation into the statistical learning of (over 270) psychology students. This formed the bulk of my PhD thesis (Gordon, 1998). More recently, my research led me to explore the ideas of university educators about pedagogy – in statistics and other areas of higher education. In this self-study, the question framing my review of research was this: what has qualitative inquiry offered to improve my understandings of teaching and learning?

In particular, a collaborative project investigating teachers' ideas about teaching and learning service statistics has enabled me to advance my thinking about teaching statistics at the Mathematics Learning Centre and helped me to re-energise my approach.

Research on Teachers' Experiences of Teaching and Learning Service Statistics

Project Outline

The investigation consisted of a three-phase series of e-mail interviews with statistics educators from around the world. Participation was invited through an electronic request to the membership list of the IASE (International Association for Statistics Education) and Australian bulletin boards. The interview transcripts from 45 respondents (over 70,000 words) formed the raw material of the study, which I will refer to as the IASE study. We have written about different aspects of the data in a series of publications. These include the teachers' conceptions of teaching service statistics courses and their ideas about what makes a "good" statistics student and a "good" statistics teacher (Gordon, Petocz, & Reid, 2007, 2009).

The participants of the IASE study were from many countries. The diverse cultural base of our respondents immediately enabled me to access a multitude of different views about effective teaching and hence opened the window to insights that would not be available by studying my own practice and that of local colleagues. In addition, by collaborating with two researchers from different backgrounds and research areas, I was able to test and amplify my own interpretations of participants' responses through readings and understandings of my co-researchers, a way of "co-knowing" (Leont'ev, 1978, p. 60).

Opportunities for My Learning Provided by This Research

This IASE study was a rich source of ideas for my self-study. Some were practical ideas that could be applied to my teaching. These ideas included strategies for helping students improve their communication skills and ways of addressing "maths phobia". Participants shared their online resources, software and presentations with us. They offered suggestions for helping students learn to interpret statistical information critically and for engaging students by what one respondent called "vivid story telling" about real research projects "including what can go wrong and how to learn from errors that other people have made".

The project enabled me to draw on the passion and commitment of experienced statistics educators. An unexpected benefit of this, and one germane to my self-study, was that the email "conversations" reminded me of the joy of teaching statistics. I will clarify this through two short excerpts from participants Johanna and Heintje, below. These were responses to our question: *What are the attributes of "good" statistics teachers?*

Johanna:

> A good statistics teacher is someone who is passionate about teaching, students, and statistics and who can combine those passions in an effective manner. A good statistics teacher is enthusiastic about her/his work and that enthusiasm shows in the classroom. When doing my own dissertation research, I looked at experts in the field of statistics education and identified a kind of "magic" that each of them had in the classroom. I am not sure how to qualify this "magic" in any other word. Passion and enthusiasm are a must. A good statistics teacher is not someone who is teaching only so that s/he can do research or just as a job – a good statistics teacher wants her/his students to learn and be excited about learning.
>
> I absolutely love teaching statistics, particularly to students in introductory statistics courses, who are taking the course because they "have to" and not because they choose to. This makes my job more challenging. I have also been considering the concept of the "magic" I talked about in the classroom and whether or not that can be measured in some tangible way. If I figure something out, I will let you know.

Johanna's views about teaching students who "have to" study statistics prompted me to reframe my thinking about students' reluctance to learn statistics. Rather than regret this aspect I could welcome it as a challenge. Further, I found Johanna's consideration of "magic" in teaching statistics to be novel. Statistics as an academic subject is not often associated with magic and inspiration. To think that I could aspire to bring magic to teaching statistics was hopeful even if problematic.

Heintje provided a different insight on good teaching.

> A good statistics teacher will stimulate students to take their own initiative, to become confident about themselves in doing statistics, exploring data, discussing subjects with other students or teachers. A good statistics teacher will help students to overcome their statistics anxiety and will take care that they get familiar with the discipline step-by-step, embedded in a psychological context. A good teacher will also be a good listener and will seriously consider student evaluations as a means to improve the educational design.

Heintje's report was a useful summary on what it could mean to focus on students' worlds. Constructively, Heintje pointed out that a student who was "perhaps less talented and less motivated" in statistics might be an excellent student in other domains.

> It might be the case that such a student knows a lot of psychological treatments or practices, is a good writer or has excellent social or communicative skills. All these (different) skills and talents might be of great importance for the group assignment or group task.

Heintje's closing observation stimulated me to think more broadly about what is needed to solve statistical problems. In professional life, someone faced with making a decision on the basis of data would be likely to employ a range of strategies, including discussing the problem with colleagues and drawing on experts. Hence providing students with opportunities to experience these strategies could be important even in early learning of statistics. I now incorporate group work on discussing "real" problems from medical research or current media reports in my statistics bridging courses and provide opportunities for students to access authoritative statistical resources (such as the Australian Bureau of Statistics website) to enable them to uncover information that interests them and critique this information.

The IASE study enabled me to engage with the international statistics education community. It stimulated me to "make" time to read the statistics education literature and hence continue to broaden my understandings about teaching and learning statistics. The study also stimulated new research about effective pedagogy. In our ongoing research project, participants are from a range of professional areas such as teacher education, accounting and engineering (Reid, Petocz, & Gordon, 2010). In summary, the ideas and initiatives of the participating statistics educators are a source of motivation and inspiration to me and help me identify new avenues of research as well as effective teaching tools and strategies.

Identifying Relevant Tensions

The IASE investigation also alerted me to tensions and dilemmas concerning statistics education at university. The first tension is between what students perceive as their real needs – their current concerns or their goals for their future lives – and the theoretical knowledge presented to them in the academic context. An issue here is the common practice of introducing students to statistics in the junior years of their undergraduate degrees, often in a decontextualised form, with only minimal connection to their home discipline.

The second tension is between high quality learning and the current conditions surrounding higher education in Australia and other countries. These conditions include increasing demands on both students and staff and an environment where staff "performance" is monitored (Buchanan, Gordon, & Schuck, 2008) and where evaluations of teaching may be based primarily on student satisfaction surveys. This environment could lead to expectations of a "pizza delivery" model of education.

That is, conditions could lead to demands for the economical transmission of information, efficiently delivered and in a form that is easy for students to consume – if not to digest.

The third tension concerns the identity of expert and learner in our information rich and technologically advanced era. Vygotsky (1962, 1978) proposed that the young develop through being instructed and assisted by the more experienced members of the community. However, expertise in our technological era is as often created by the young and relatively inexperienced as by those more generally accepted as experts. There is also a mismatch between the modes of seeking and communicating information that are familiar to most students, such as Google, Facebook, Twitter or eBay, and the less technologically based methods of teaching statistics that are still commonplace in many universities, such as large group lectures and textbooks.

These tensions are in the wide arena of higher education and society and cannot be resolved by an individual. However, my awareness about the broader institutional context assists me to organise my practice more effectively within organisational mechanisms and to appreciate democracy and collegiality at the more local level of my unit.

Opportunities to Learn Through Experiencing Incongruity in Students' Actions

My research about teaching and learning statistics indicated that helping students relate statistical concepts to their personal situations is of the utmost importance. In order to align my "teaching intent with teaching actions" (Loughran, 2007, p. 12) in daily practice, I implemented activities that would encourage students to link statistics with their everyday lives and understandings. This led to experiences of incongruity as the following examples demonstrate.

One activity was to ask students to construct metaphors to describe statistical concepts. A group was working on confidence intervals, including ways of calculating them. I asked them to think of a metaphor for a 95% confidence interval for a population mean. After discussion, a few different metaphors were offered. One student mentioned trawling for a fish – casting a net into the sea. I was very pleased with this metaphor and pointed out its strength: it highlighted that the 95% probability was not associated with the whereabouts of the fish but with the ability of the net to capture the fish. The fish is where it is (as is the population mean), but the net is cast based on a sighting of fish (analogous to a sample mean). The students digested this idea in silence and then one voiced this opinion: "I hate fishing!" There was a chorus of agreement. And the metaphor immediately became unpopular and unusable. The challenge to me was that although the student's comment seemed absurd and irrelevant, it resulted in the failure of my mission to debunk a common student misconception. More deeply, it was a jolt to my self-complacency in thinking I had solved the problem of engaging students in thinking deeply about this important statistical concept.

A second example of using metaphors concerned hypothesis testing. I generated an analogy to explain the concepts of the null hypothesis and the alternative (research) hypothesis as follows. The null hypothesis is like the presumption of innocence. It is up to the researcher (prosecutor) to provide evidence against this presumption in order to get a guilty verdict (conclude the alternative hypothesis). Immediately one student cried out: "I cannot think of myself as a prosecutor!" Here, my desire to link statistics to what is meaningful to students was too successful – by internalising the analogy, connecting it to her beliefs and life philosophies, the student rejected the representation.

These incongruities awakened me to realisations about "authenticity", not in the narrow sense that the statistical learning tasks correspond to the so-called real world, but in the wider sense that the "student's being is engaged in the teaching–learning interaction" (Kreber, 2010, p. 191).

A further incongruous example of a student engaging personally with statistical theory occurred when Anne expressed her concerns about the use of statistics. She said this. "When I see a normal distribution I become concerned for its use. Who's fitting what into it for what purposes? Who benefits?"

The idea of fitting something or someone into a normal distribution invoked in me a vision of placing a two-dimensional human form into a mathematical representation: the graph of a probability density function. However, this strange vision did alert me to two aspects that require care in my teaching. One is that students learning introductory statistics may not distinguish between a graph of a frequency distribution showing a finite set of (actual) data points and a theoretical or mathematical model of a phenomenon (such as the normal distribution of a variable). Secondly, and most importantly, there is a need to consider ethical dimensions of statistical analysis and interpretation. These are often totally ignored or pushed into the background in a statistics course. Hence the incongruity within this student's passionate declaration taught me some lessons, which I now endeavour to heed.

An entirely different experience of incongruity occurred when a distressed student in my statistics tutorial group asked to see me in private to discuss the crisis he was experiencing in trying to study statistics for psychology. This student had not studied mathematics at school and was completely at sea with the fast moving and complex lecture material. He described to my colleague and me his insurmountable difficulties with learning the material and lamented that his attempts to cope – so clearly doomed to failure – were breaking his heart. I gave this student two "good" pieces of advice. One was to drop the subject before census date (the date on which the subject would be recorded on his academic transcript). The second was to continue to study the subject at his own pace – to work through the lecture notes, attempt the exercises and come to the Mathematics Learning Centre for help as usual. Then continue this informal study during the vacation, working through the content a second time, if possible, and re-enrol for the subject next year.

To my chagrin both these pieces of advice were ignored. The student kept up his enrolment and continued his ineffective efforts to learn little bits of the curriculum, like snatching statistics parts off a moving conveyor belt. What was incongruous was that after our talk his demeanour changed dramatically. He worked cheerfully and positively, contributing (usually wrong answers) to our small group statistics workshops in the Centre and evidently no longer daunted by the Herculean task. (He also expressed huge appreciation to me afterwards for my teaching.) I conclude that by not taking my advice the student came to see that he, as a person, was not being judged and found wanting; rather that he was simply engaged in a course of action that he could choose to pursue or not.

Finally, reflecting back on my reactions to Norman's frank comments about his aims, outlined earlier in this paper, I realised that what was unsettling to me was an incongruity in his situation. There was a mismatch between Norman's expressed surface approach to learning statistics and his interest in mathematics at the level of philosophy and abstract thinking.

A further insight has emerged from all these reflections. Incongruity can be a signal in self-study, a jolt to "notice" (Mason, 2002) or become aware of a situation or problem in teaching. Loughran (2006) points out that noticing a problem immediately brings it into focus and so makes possible a response. If a teacher is focussed on her (or his) own "performance", she may not notice her students' lack of engagement or understanding. I claim that incongruity can provoke awareness and with this awareness the potential to make changes. Hence being receptive to experiences of incongruity in students' actions and our own is useful in self-study.

Synthesis: Reframing Practice – Educating Statistics Students I Teach

According to Davydov and Markova (1983, p. 57), for educational activity to develop: "it is necessary to ascertain and create conditions that will enable activity to acquire personal meaning, to become a source of the person's self-development". This self-study provokes the question: What sort of self-development could I hope that my students will achieve? The students are, after all, fully developed adults by the usual criteria.

There are, it seems to me, three areas in which I can aim to promote students' development: intellectual development, the development of students' conceptions of statistics and their personal growth.

Intellectual development is characterised by an improved capacity for abstract thinking, better methods of learning, and conscious control over the processes of learning. Educational activity in this context would be made up not only by those actions aimed at the mastery of knowledge, skills and technical abilities but also by those directed at enhancing mental capacities such as the ability to reflect, to understand the connections between concepts and to articulate and communicate statistical ideas.

It could also be expected that educational activity would result in students developing their conceptions of statistics. My research and that of colleagues has shown that undergraduate students in service courses have a range of conceptions of statistics. Some conceived of statistics as totally meaningless (or even a form of academic oppression), some as decontextualised formulae and algorithms or as irrelevant information to be mastered and stored in the short term to meet the demands of assessments, while a few showed awareness of statistics as a tool for future professional work and, ultimately, for thinking critically or making sense of the world (Gordon, 2004; Petocz & Reid, 2005). The question of how to provide a learning environment that orients students toward the more extensive conceptions of statistics is an ongoing challenge.

A further and important dimension of educational activity in statistics is to ensure that students are aware of their ethical responsibilities to use statistical knowledge correctly. A respondent in the IASE project expressed this poignantly in the excerpt below (translated from his Spanish original).

César:

> I always indicate to my students that lies are bad in themselves, but when they are based on statistics they are doubly bad, because they carry the power of conviction of numbers and sophisticated methods (which are esoteric and mysterious for many people). The idea is to provoke in students the ethical commitment that the statistical task requires. That for this it is necessary to acquire a thorough grasp of statistical theory, to know its extent and limitations, to be able to apply them correctly in practice and, if they have any doubts on the matter, to go to those who will know how to solve the problem correctly: a good book, a professor, a colleague.

Finally, and perhaps most importantly, an area in which an adult learner can develop by studying a statistics course is that of personal growth. Students who succeed in overcoming their reluctance to tackle mathematics based courses, who conquer long-standing difficulties with mathematics or a severe lack of confidence in their abilities to do mathematics, often report feelings of achievement such as are reported below by two students.

Jane:

> I'm glad that I'm doing maths, 'cause for me that has a lot of value. Maths has kudos for me. I found that I actually knew more than some other people. Instead of being the one who knew the least. And that gave me an enormous sense of power. Or a bit of power anyway, that I could actually know more than somebody else.

Summing up her feelings, Sandra, after successfully completing her statistics course, reported that:

> All my life it felt like I had this dark secret: – that I felt really stupid about this area. I'd cover it up so no-one would know. This really feels like growing up.

As the above two reports indicate, students can find learning statistics empowering.

Conclusion

In this self-study I have selected highlights that illustrate the uncertainty and ongoing disarray that characterise my practice. Reflecting on these selected dimensions of practice has showed me the influence of and inspiration provided by the students I teach and the colleagues and participants in research and scholarship projects. Being open to incongruous events has alerted me to the inherently contradictory and complex ways of teaching and learning. Self-study does not provide a prescription for solving problems. However, it has helped me reframe practice by adopting a state of watchfulness over my own actions as well as those of students. This means I continually question what I am trying to achieve, check whether I am providing an environment for students' educational activity (Davydov & Markova, 1983) and try to ensure that students' affective concerns as well as ethical issues are brought to the forefront. In particular, if I am "holding forth" – explaining or lecturing to students – the question of whose needs are being addressed by this endeavour is a restraining refrain in the background of my mind. While my aim is to help students progress, my own development is inseparable from theirs.

References

Buchanan, J., Gordon, S., & Schuck, S. (2008). From mentoring to monitoring: The impact of changing work environments on academics in Australian universities. *Journal of Further and Higher Education*, *32*(3), 241–250.

Davydov, V. V., & Markova, A. K. (1983). A concept of educational activity for schoolchildren. *Soviet Psychology*, *21*(2), 50–76.

Gordon, S. (1998). *Understanding students learning statistics – an activity theory approach.* Unpublished doctoral thesis, The University of Sydney, Sydney. Available online: Australian Digital Theses: http://adt.caul.edu.au/

Gordon, S. (2004). Understanding students' experiences of statistics in a service course. *Statistics Education Research Journal*, *3*(1), 40–59.

Gordon, S., Petocz, P., & Reid, A. (2007). Teachers' conceptions of teaching service statistics courses. *International Journal for the Scholarship of Teaching & Learning*, *1*(1).

Gordon, S., Petocz, P., & Reid, A. (2009). What makes a 'good' statistics student and a 'good' statistics teacher in service courses? *The Montana Mathematics Enthusiast*, *6*(1, 2), 25–39.

Kreber, C. (2010) Academics' teacher identities, authenticity and pedagogy. *Studies in Higher Education*, *35*(2), 171–194.

Leont'ev, A. N. (1978). *Activity, consciousness, and personality* (M. J. Hall, Trans.) Englewood Cliffs, NJ: Prentice–Hall.

Leont'ev, A. N. (1981). The problem of activity in psychology. In J. V. Wertsch (Ed.), *The concept of activity in Soviet psychology* (pp. 37–71). New York: M. E. Sharpe.

Loughran, J. (2006). A response to 'Reflecting on the self'. *Reflective Practice*, *7*(1), 43–53.

Loughran, J. (2007). Researching teacher education practices: Responding to the challenges, demands, and expectations of self-study. *Journal of Teacher Education*, *58*(1), 12–20.

Loughran, J., & Northfield, J. R. (1998). A framework for the development of self-study practice. In M. L. Hamilton (Ed.), *Reconceptualizing teacher practice: Self-study in teacher education* (pp. 7–18). London: Falmer Press.

Marton, F. (1988). Describing and improving learning. In R. Schmeck (Ed.), *Learning strategies and learning styles* (pp. 53–82). New York: Plenum Press.

Mason, J. (2002). *Researching your own practice: The discipline of noticing*. London: Routledge Falmer.

Nicholas, J. (2010). *Mathematics learning centre annual report 2009*. Sydney: The University of Sydney.

Petocz, P., & Reid, A. (2005). Something strange and useless: Service students' conceptions of statistics, learning statistics and using statistics in their future professions. *International Journal of Mathematical Education in Science and Technology*, *36*(7), 789–800.

Reid, A., Petocz, P., & Gordon, S. (2010). University teachers' intentions for introductory professional classes. *Journal of Workplace Learning*, *22*(1, 2), 67–78.

Samaras, A. P., & and Freese, A. R. (2009). Looking back and looking forward: An historical overview of the self-study school. In C. A. Lassonde, S. Galman, & C. Kosnik (Eds.), *Self-study research methodologies for teacher educators* (pp. 3–19). Rotterdam, The Netherlands: Sense Publishers.

Schuck, S., & Russell, T. (2005). Self-study, critical friendship, and the complexities of teacher education. *Studying Teacher Education*, *1*(2), 107–121.

Vygotsky, L. S. (1962). *Thought and language*. Cambridge, MA: The MIT Press.

Vygotsky, L. S. (1978). *Mind in society*. Cambridge, MA: Harvard University Press.

Vygotsky, L. S. (1981). The genesis of higher mental functions. In J. V. Wertsch (Ed. & Trans.), *The concept of activity in Soviet psychology* (pp. 144–188). Armonk, NY: M. E. Sharpe, Inc.

Chapter 10
Reconstructing Teachers of Mathematics

Peter Pereira

Dilemmas of Teaching

When I first started teaching teachers, after 10 years of secondary school mathematics teaching, I had a relatively simple view about what my task required. I knew that improving instruction involved teachers' understandings about the teaching/learning process as well as their understandings about content and materials. But I did not fully realize that mathematics teachers, and *their* teachers, also need to develop greater understanding about themselves as learners and teachers. As a result, I tended to focus on content knowledge and on pedagogical knowledge. From this perspective, the goals of teacher education, though difficult to achieve in practice, are easy to describe: make sure that teachers have the necessary content knowledge and then, as time allows, provide them with as much pedagogical knowledge as one can.

I quickly found that this relatively simple picture of mathematics teacher education was not adequate. Although teachers in my courses seemed to have mastered a lot of content and they certainly knew procedures for teaching it, these gains did not seem to affect their practice. Novice teachers, when faced with the need to survive in schools, fell back on what they had experienced when they were students. Experienced teachers kept on doing what they were comfortable in doing. Clearly the process of becoming a mathematics teacher, especially one committed to reforming mathematics teaching, was more complicated than I had thought.

The complications stemmed from two sources. First, the distinction between just two kinds of teacher knowledge—content knowledge and pedagogical knowledge—is too superficial. As discussed in Chapter 1, Shulman (1986) argued that this distinction is both unclear and incomplete, inventing the term "pedagogical content knowledge" to indicate what was usually left out. Although contemporary researchers may differ about the meanings of the various terms involved, most agree

P. Pereira (✉)
School of Education, DePaul University, Chicago, IL, USA
e-mail: ppereira@condor.depaul.edu

S. Schuck, P. Pereira (eds.), *What Counts in Teaching Mathematics*, Self Study of Teaching and Teacher Education Practices 11, DOI 10.1007/978-94-007-0461-9_10,

that teacher knowledge is a complex concept that goes beyond traditional distinctions between content and methods. The second complication arises because any learning process inevitably includes both cognition and affect. While this has been recognized for a long time, few researchers have studied teaching and learning as more than intellectual activities. Although we may be able to study them separately, in practice cognition and affect are inextricably linked.

I learned this vividly during a workshop in which we deliberately combined reflection on the emotional aspects of learning with the study of mathematics content. From the outset, we operated on the premise that improving instruction was complicated because it involved teachers' understandings about content and materials, about the teaching/learning process, and about themselves as learners and teachers. Consequently, we attempted to make changes in all three kinds of understandings, balancing the development of content knowledge with reflection on learning and on self. To do this, we all wrote stories about our experiences as learners, and we set aside time for participants, and ourselves, to discuss our emotional experiences as teachers and learners. Both stories and discussions combined cognition and affect and were graphic, immediate and personal, enabling us to come to grips with the concrete particulars of our experience.

As a result of these experiences, as well as of my reading of research, I now view teaching as full of objective dilemmas and subjective tensions. Should I intervene or should I allow teachers to figure things out for themselves? Should I try to reduce frustration or is a certain amount of frustration a necessary part of learning? Should I tell the whole truth or should I withhold some information? How should I respond to a student's evident need for praise? There are no general answers to these questions. Teaching requires us to make judgments and take practical action in an uncertain context. Thus these dilemmas are both genuine and unavoidable. They are *genuine* because they cannot be resolved in advance by following a prescribed course of action, and they are *unavoidable* because they will always be with us whenever we teach. Much as we might like to have rules to tell us how to resolve them, we have to learn to live with them. Indeed, negotiating the pitfalls and possibilities can make teaching exciting, both for teachers and their students.

In an earlier paper (Pereira, 2005), I wrote about some of these dilemmas, the tensions I experienced, and how I negotiated them. In this chapter, I will describe dilemmas and tensions that arose as I helped practicing teachers to reconstruct themselves as teachers of middle school mathematics.

Experiencing Mathematics in a New Way

The reform movement in mathematics education has been partly driven by results from national and international assessments of mathematics knowledge indicating that many students, especially in the United States, were not learning enough mathematics. The most widely publicized of these studies, the Third International Mathematics and Science Study (TIMSS), compared the mathematics achievement of students from 41 different countries (Schmitt, 1997). In general, students in the United States ranked in the middle of this group, well behind most industrialized countries and certainly well below Singapore, the top performing country. Though there are many ways to interpret this research, it was generally seen as indicating a need for improving classroom mathematics teaching in the United States (Stigler & Hiebert, 1999).

This impression was significantly reinforced by the publication of a book by Liping Ma, an American researcher fluent in Chinese, who interviewed 23 mathematics teachers from the USA and 72 mathematics teachers from China (Ma, 1999). In her interviews, she probed teachers' grasp of mathematics and issues surrounding their teaching methods. In both respects, the Chinese teachers seemed to be well ahead of their counterparts in the United States. For the most part, the American teachers could perform basic calculations, and generally they could explain a technique for getting the solution. On the other hand, they did not have what Ma called a "profound understanding of fundamental mathematics" (PUFM). An alarming number of the US teachers did not seem to have a conceptual grasp of the problems they were given. In contrast, the Chinese teachers not only could grasp the conceptual underpinnings of the questions, they were often willing to go well beyond the question asked, even providing multiple solutions on occasion. These differences carried over to questions about teaching methods where, for instance, the Chinese teachers were able to show how to model the central mathematical concepts with manipulatives.

A common response to these kinds of results, especially from mathematicians but also from the general public, is that teachers do not know enough mathematics and therefore have to learn more content. There is plenty of evidence to support this view, both from research and our own personal experiences. Many mathematics teachers do not have an adequate knowledge of the subject they are teaching. Still, it is a complex problem (Ball, 1988; Ball, Thames, & Phelps, 2008), and one has to question the apparently obvious premise that a better understanding of mathematics content would result in more effective teaching of mathematics. It certainly is necessary to improve the subject matter preparation of mathematics teachers in the United States, but is it sufficient? Both the TIMSS study and Ma's work indicate that US teachers have had about the same amount of formal mathematics training as have teachers in other countries. Yet their students do not perform as well. Moreover, there is little research that links mathematical content knowledge (MCK) with what actually takes place in mathematics classrooms.

As one looks more carefully at the recommended directions of reform, it becomes increasingly evident that not only are we asking teachers to learn more mathematics, we also are asking them to change their understanding of what mathematics is. Reform recommendations move us away from a view of mathematics as only a collection of skills, techniques, and concepts toward a more comprehensive view of mathematics that includes a set of identifiable processes and habits of mind. Five such processes are specified in the *Standards* (NCTM, 2000) and the *Curriculum Focal Points* (NCTM, 2006). Although students are still expected to acquire specific mathematical knowledge and to master designated skills (as specified in the *content* standards), they are also expected to engage in mathematical activity (as specified in the *process* standards). To see how unfamiliar this might be to some, consider what is implied by the process standard on problem solving. Students are now expected to think through situations where there may be no routine route to a solution, or at least no route known to them and perhaps not one known to the teacher. Each student is encouraged to use an individual style of problem solving and to approach the problem as they think best. Problems often do not have one correct answer, but many. The teacher's job is not to get the student to the one right answer or to use the correct algorithm. Instead, the teacher is expected to focus students on the process of problem solving, in its various appropriate forms, so that students are better equipped to tackle problems in the future.

All of these processes involve habits of mind. Paul Goldenberg (1996) identified some of them (e.g., tinkering, looking for invariants, visualizing, reasoning by continuity, going to extreme cases), and these were included in geometry materials for secondary students (EDC, 2000). In a recent article, he and his colleagues argue that mathematical habits of mind can be used as an organizing framework for most mathematics instruction (Cuoco, Goldenberg, & Mark, 2010). This framework nicely captures what most reform mathematics educators would like to do. Still, mathematical habits of mind do not come easily to students, especially to those who may have developed other, competing habits. Take, for example, the habit of generalizing. Once a problem is solved, a mathematically inclined person (i.e., a person with well-developed mathematical habits of mind) would pause to reflect, either to look for a different solution or to seek a more general solution. On the other hand, most teachers would be quick to move on to the next problem; after all, there is a lot to "cover." Or, to take another example, suppose teachers were asked to find the maximum area of a rectangle of fixed perimeter. What would come naturally to them? Many might search for a formula or an algorithm; others might try to evade the question. Some, we hope, would almost automatically go to an extreme case: a long, skinny rectangle with virtually zero area.

Unfortunately, many mathematics teachers have never experienced mathematics this way. Problems to them are usually "one-step" problems where the only two numbers in the problem must be combined. Thus the only task for the student, once the two numbers have been identified, is to figure out whether to add, subtract, multiply or divide. Having seldom experienced genuinely problematic situations in mathematics, they tend to see problems as having only one solution. Is it any wonder then that they assume there is only one approved method for reaching that solution? If they have only experienced mathematics as the acquisition of procedural knowledge, they will have a hard time when they are expected to help students develop conceptual knowledge, especially conceptual knowledge of their own meta-cognitive processes. Liping Ma's work underscores the depth of this gap. In her view, teachers need to develop profound understanding of fundamental mathematics, something not easy to obtain when one has been subjected to 12 or more years of mathematics as skills, techniques, and algorithms. It takes more than courses in "mathematics as usual" to overcome the damage caused by many years of such instruction. Teachers need to be reacquainted with who they are as learners of mathematics so they can experience mathematics in a new way. In short, they need to reconstruct themselves as learners of mathematics before they can reconstruct themselves as teachers of mathematics. Still, as we shall see in the next section, there are obstacles along the road.

Becoming Learners of Mathematics

In the spring of 2010, I taught a middle school geometry course to practicing urban teachers, most of whom were teaching in middle schools and some of whom were already teaching mathematics. The students were enrolled in a master's degree program designed to prepare them to be middle school mathematics specialists. A few had a good grasp of mathematics, but in general, there were gaps in what they knew, conceptual misunderstandings, and inappropriate mathematical habits. Typically they saw mathematics as an unrelated collection of skills, techniques, and verbalizations. Still, all of them had been judged to be successful teachers, and all of them genuinely wanted to make a difference in the lives of children.

The program as a whole challenged teachers to develop mathematical ways of thinking and to value mathematics as an activity. In this course, which came early in the program, I wanted these teachers to begin to see themselves as learners of mathematics (who can solve problems, reason mathematically, communicate their thinking, make connections, and represent mathematical ideas in a variety of ways) with mathematical habits of mind (which value invention, conjecture, explanation, and argument) and reflective habits (which encourage them to examine their personal experience—past and present—in order to understand the interactions that guide their learning). For most, if not all, this meant that they had to make significant changes in order to experience mathematics in a new way. As you will see, they had—and some still have—many powerful emotions and ingrained habits to overcome.

Fears

As part of this course, I asked the teachers to keep a journal in which they described their emotional experiences—both past and present—when learning mathematics. At the beginning of the course, the most common emotion expressed was fear—fear of failure, fear of exposure, fear of inadequacy, fear of humiliation, fear of intimidation. Comments like the following abounded: "I entered this class fearful, fearful of being exposed and embarrassed" or "Math scares me, I wonder if I should teach it" or "I am petrified by math" or "I dread the thought of embarrassing myself in front of my students." This kind of comment far outweighed the more neutral comments ("Math is uninspiring" or "just a bunch of rules to be memorized"). Comments suggesting that mathematics might be interesting, exciting, or even beautiful were rare. If this is how practising teachers felt about mathematics, how could they teach it? They needed help to overcome their negative feelings.

This should come as no surprise to anyone who has taught mathematics courses to teachers. Much anxiety and fear surround mathematics, especially amongst elementary and middle school teachers. So we ask them to keep journals, we affirm their feelings, we try to create supportive and trusting environments, and we encourage them to experiment without fear of failure. Still, although I did all of these things (and, if students are to be believed, successfully created a different culture of learning), it is a time-consuming process, especially with adults who have had many years to insulate themselves. Some teachers found it easy to become more aware of their feelings as they learnt mathematics; others were more reluctant, perhaps for lack of practice or because feelings and mathematics did not seem to go together. But for all, as was made clear by their comments, their fears, even when recognized, were hard to overcome. As one teacher summed it up:

> Teachers are funny creatures. We constantly put our kids on the spot and try to challenge them to break out of their shells, take risks, and encourage them to believe that even if you are wrong (publicly) it is okay because you still learned something. But when the tables are turned on us—look out—we all duck for cover!

What made it so difficult for these teachers to let go of their deep-seated emotions about mathematics? Why was it not enough to recognize and affirm their feelings? As one teacher pointed out, letting go requires us to give up something.

> Even after taking this course, I am not sure that I am able to let go of [my] fears. All my years of schooling and teaching have left behind too many patterns of behavior that are hard for me to change. How can I give these up when I have nothing to replace them with?

If, as is true in many cases, we must give up something that has helped us to cope—a habit that brings order to a confusing world or a belief that has become part of who we are—this can be a painful and sometimes lengthy process. It is, however, an essential step if one is to become a mathematics learner.

Frustrations

As the teachers were challenged to develop more mathematical ways of thinking, frustration often arose. Indeed, at one time or another every journal mentioned frustrating incidents. Terry wrote:

> Frustration was another thread woven throughout my journals, frustration at my own limitations and misunderstandings about geometric concepts. I was often frustrated at what I couldn't 'see', especially when others appeared to grasp concepts so easily.

Unfortunately, for some this was a common occurrence. As one teacher informed me, "The majority of the time I experienced feelings of frustration, irritation, and aggravation." While aggravation was not my intention, when teachers were able to work through their frustrations, they became invested in the outcome and meaningful learning took place. For example, Ivy wrote:

> When we began the tangram activity, I played with the pieces for a while, and was making no headway. I was moving pieces around one at a time trying to make a polygon with all

> seven pieces, and was getting frustrated. All of a sudden, I looked down to find that I had made a parallelogram.

She then goes on to describe the flurry of activity and insight that ensued as she and her group discovered all 13 ways to make a convex polygon out of tangrams and developed an argument that there were no other possibilities.

Still, frustration can end up disempowering the learner. This can be an overwhelming experience, especially when one is not used to it or, even worse, when one has never experienced the elation that comes after having successfully gained insight by means of one's own efforts. All too often this was the case for many of these teachers who, regardless of social class or ethnic background, had been deprived of such experiences. Given that these are a foundation for becoming a mathematics learner, I found I had to walk a tightrope. Providing answers too soon could deprive teachers of important and necessary experience, providing them too late could allow their frustration to spiral out of control.

A certain amount of frustration is an essential part of any valuable learning experience. It is important for mathematics learners to understand this and to be able to overcome their frustrations. It is also important for them to accept the fact that no matter how much we see; there is always something more to be seen. As Amy put it:

> The sense of frustration of knowing I probably wouldn't have found it on my own is enormous. The ... video is another example of someone or many someones 'seeing' something and finding the elegance of it. Is it that their brains work differently? Is it that they have had so much experience working with these figures and concepts that 'seeing' is simply the next step in the evolution?

"Seeing" is the result of some hard work and some dry spells will inevitably occur. Learning how to reduce such spells by coping with frustration, ambiguity and uncertainty is another essential step in becoming a mathematics learner.

From Learner to Teacher of Mathematics

So far I have separated the process of becoming a learner from the process of becoming a teacher, appearing to suggest that it was possible to do one first and then the other. But I found that the situation was more complex than that. For one thing, these teachers were teaching their own classes, so it was only natural for them to look for connections between what they experienced and how they taught. For another, while we can separate teaching from learning for analytical purposes, in practice the two get mixed together. Teachers' experiences when learning and their experiences when teaching were intertwined, as the following comment from Julie's journal illustrates:

> I am overcoming my frustrations and learning to explore; and I am trying my best to adopt your method of keeping your mouth shut and letting us struggle. It is so hard for me to see my students struggle or become frustrated when solving a math word problem. I want to help them.

Julie was talking about her frustrations, about what she had seen me do, and about her attempt to teach differently all in a few sentences. This was a common occurrence. Like Julie, most teachers found it difficult to talk about their learning without also talking about their teaching. It became clear that it is impossible to *first* reconstruct oneself as a learner and *then* reconstruct oneself as a teacher. One has to do both at once.

The situation was complicated by the fact that teaching has a subjective side as well as an objective content. Our fears, our frustrations, our hopes, and our actions have an internal dimension. They are not just a reaction to external events. They are mixed with the emotions, wishes, and fantasies—derived from life histories, recent events, relationships with others, and prior classroom experiences—which we bring with us. Any curriculum designed to reacquaint teachers with who they are as learners and teachers of mathematics will inevitably expand beyond the intellectual aims of the objective content to include emotions inherent in the teaching and learning process. Much as we might try to ignore them, they are there and need to be recognized. As with learning and teaching, in practice cognition and affect also get mixed together. Attention to the dynamics involved, and the meanings created, can help teachers to avoid pitfalls and maximize possibilities so that they can concentrate on the cognitive activities of the classroom.

Consider Julie's comment above. Notice that she also talks about her internal needs and desires when she says, "I want to help them." And, lest you think I have gone too far, she later confirms this possibility when she says:

> I start off by providing my students hints or providing them with questions that will lead them to self-discovery. But sooner or later I am providing them with the answer because I do not want to see them struggle. Besides I am ready to move on to the next lesson. How can I motivate my students to work on solving problems, and not just sit, give up and wait for me to assist them?

Julie likes her students and wants to help them. Moreover, she remembers how she felt. As she later asks: "Why then should I put kids in a position where they would feel powerless to learn?" This emotion was both powerful and common amongst these teachers. As Josefina said:

> I want to help my students, I remember how I felt [when I was a student], and I do not feel comfortable when I see them struggle. Moreover, they know this and are adept at getting me to help them too soon.

If, as often happens, our actions make us feel better, we are likely to continue doing them even if we recognize that in the long run they do not help students. To avoid this trap, I encouraged these teachers to become as aware as possible of their own feelings and motivations lest they trip over them when least expected. Josefina did this by recognizing that her own emotions sometimes got in her way and anticipating that sooner or later she might find herself providing her students with "the" answer before it was necessary. Sooner or later we all will, but knowing what is possible helps us to guard against this eventuality.

"Control" was another issue that loomed large in the teachers' journals. Again this has both subjective and objective components. When Carmen said, "I needed to feel in control. . . . [I now realize that] the level of inquiry (questioning) between students was always guided by me in one way or another," she was talking about her internal need and also about the way this need had influenced her practice. She went on to say:

> While I realize that "letting go" in this type of learning environment takes practice, and that students need to be "trained" on how to have class like this, it is still a scary prospect to give the impression of giving up control over the class.

The feeling that reform methods require teachers to give up control was widespread and powerful. As teachers, they were used to being in charge of what goes on in their classrooms. They were accustomed to doing this by directing students' learning, managing the flow of information, and visibly judging what is correct or appropriate. Now they were being asked to change, perhaps abandon, these practices. Not surprisingly they felt this as a loss and worried that they would lose control. Sharing my view that reform methods still keep the teacher in control does not change the depth of their feelings.

They were also being asked to change their beliefs about learning mathematics by giving up some well-entrenched habits. As learners, they expected *their* teacher to structure their learning, tell them what they should know, and evaluate their responses. When, as often happened, I did not do these things, they found it unsettling. They also began to doubt, as the following comment from Sam illustrates:

> I don't know why you never answer my questions. You have to have a strong base of knowledge to gain anything out of self-directed activities. If I gave [my students] a paper plate or geoboard, I don't know what I would get from them.

Although my response ("Try it; you might be surprised.") did in fact help Sam over this hurdle, it would not be an appropriate response for everybody. It is easy for us—who are ourselves in control—to underestimate the power of these subjective feelings. To move beyond them, we have to get teachers to think about what they can and cannot control. So I kept asking questions. Why do we have this need to feel in control? Why do we have so much trouble in giving it up? What do we think we are in control of? What do we need to be in control of? We cannot change our needs to be in control of things; these come naturally to all of us. But we can change *what* we aim to control and *how* we go about it, thereby replacing what was lost with a more powerful vision.

While this may seem complicated, there is good news. Because the dilemmas and tensions the teachers experienced when teaching were similar to the dilemmas and tensions they experienced when learning, we did not have to look at each activity separately. Focusing teachers on how they learned in the geometry class also helped them to think about their teaching. When, for example, Sam expressed his concern about my behavior in the geometry class, he was also grappling with his own behavior when he taught. Josefina's reluctance to let her students struggle stems from the same sources as her own discomfort when she was asked to explore, conjecture, and discover. Carmen's need to feel in control applied to her learning as well as to her teaching. Whether the result of intra-personal, inter-personal, cognitive, or pedagogical factors, the dilemmas and tensions that arose in one context were likely to crop up in the others. For this reason, I will look at some of the dilemmas and tensions I experienced in the geometry course.

Teaching Mathematics Teachers

Many teachers introduce a new topic with lists of vocabulary for their students to memorize. I almost always reversed this order by asking the teachers to mentally or physically construct something before naming it. There seemed little point in having them name objects when they did not have any clear idea of what they were naming. So, for example, when we were studying various kinds of polyhedra, I did not focus on the names of these polyhedra ('cuboctahedron' for example) unless we needed the name to communicate something to someone else. After all, we could easily look up a word when we needed to. Yes, teachers need to learn vocabulary, but there is no point in doing this just to be "correct." They should do it to communicate their thinking to others and to themselves at a later time. In the geometry course, the experience came first and the definitions later, once a concept of what was being defined was clear.

Still, this is not the normal pattern of instruction, a pattern often reinforced by many years of experience as a student. As a result, these teachers got uneasy when they were asked to sort out their ideas before naming anything. There were times when it appeared to them that I was withholding important information. "If you know the name, why not just tell me?" was a frequent comment in the journals. At such times their internal tensions built up, sometimes to disabling proportions as they began to doubt the trustworthiness of their teacher. As their tensions built, so did mine. Although I certainly wanted to help them, I also wanted them to develop different habits. Should I refuse to give them the information they sought, risking an increase in doubt and mistrust? Or should I give in to their demands and miss an opportunity for mathematical growth? There is no "right" answer to such questions; we make our judgments in individual cases.

Once having constructed something—such as a tessellating pattern or a three-dimensional model—there was an object, in which they presumably had an investment and some interest, to talk about. I encouraged teachers to describe their constructions in as precise a way as possible. Although I did not expect them to use formal terminology, I did want them to take notice of what they had done and then describe the object in some detail. This was hard for many of them to do, especially if they were asked to do it in writing. For one thing, they lacked the vocabulary, not just the names of things but more importantly vocabulary to describe mathematical relationships. More serious, they often did not feel a need for general and precise description. It seemed enough to have made the object and to appreciate what they built. When pushed they tended to say "you know what I mean." In many cases perhaps I did, but did they?

Consider the following situation. Raul is trying to see what 3-D models he can make if he only uses squares, a worthy goal but one with limited possibilities. At one point some equilateral triangles were thrown into the mix, and Raul created an object he found far more interesting. Here is what he had to say when pushed to write about his creation:

> I ran out of squares, so I included some triangles with the squares I already had. This substitution gave me a more interesting shape.

End of story. No further description of this interesting shape; no mention of what made it interesting. So we are left to guess about what it actually looked like or what Raul noticed that made it more interesting. Yet I knew he had created a semi-regular polyhedron—a model of the Archimedean polyhedron called a cuboctahedron. Moreover, I thought I had heard him talk, albeit vaguely, about the object's symmetry, the imaginary surrounding sphere, the fact that two triangles and two squares came together at each vertex, and even that the triangles and squares were regular polygons. Or maybe I imagined it; we have no record of what he actually said and he seemed unable to put much in writing. As a consequence, the important mathematical content of a potentially powerful experience got lost, thereby lessening its potential to affect his practice in the future.

As you can imagine, I was disappointed. Moreover, this was not just an external reaction; my internal tensions increased as well. Often this happened, before I fully recognized the situation. So frustration, uncertainty, and sometimes anger also were involved. How, I wondered, could teachers, such as Raul, be so unobservant, have so little desire to be precise, or miss opportunities to generalize? So, since these teachers had not done it for themselves, I was often tempted to tell them what they had built, what they had seen, and what they "ought" to know. In short, I sometimes felt compelled to take over the process, thereby losing sight of the goal that these teachers should experience mathematics in a new way by constructing knowledge for themselves. I was in danger of succumbing to urgent but more dimly perceived motivations.

Fortunately, in this case I was able to avoid this danger and negotiate a course between competing demands. As I thought about it, I realized that Raul and I were working on different tasks. He focused on his particular construction, not on polyhedra in general. He did not see his construction as an illustration of a more abstract concept but as an object in itself. As a result, the object he was thinking about—the physical model in front of him—was quite different from what I was thinking about—the abstract concept of a *cuboctahedron*. This was a common occurrence with these teachers. In their view, the constructions they had in front of them were not related to the mathematical abstractions I had in mind.

The tensions I experienced increased still further when I asked the teachers to justify their methods of construction. Usually they were able to provide some justifications, but in general their reasoning was visual and empirical based on observation and measurement. To them, this seemed sufficient. For example, Linda had made a particularly interesting and beautiful tessellating pattern, but when I asked her to explain to me why a piece of it was a rectangle, she replied:

> It has to be a rectangle because I measured it. All the angles are right angles, and the opposite sides are equal. Doesn't that make it a rectangle?

Yes, it does mean that the object in front of her, her tessellation to which she was so attached because she herself had made it, contained a rectangle. But that was not really the question I was asking her. I wanted her to justify her construction methods, to explain why the piece *must* be a rectangle because of the way she had made it regardless of her accuracy. She was satisfied with measuring it to find out. In general, asking the teachers to make logical connections and short deductive chains seemed to them to be difficult, abstract, and unnecessary. All too often, the forms of reasoning I expected from them were different from the ones they provided.

These three sources of tension—the use of vocabulary, the nature of the objects being constructed, and the forms of reasoning expected—are examples of ways these teachers fell short of my expectations. In the geometry course I was interested in developing mathematical ways of thinking and habits of mind. Still, I could not escape my belief that mathematics also requires knowledge of content and mastery of technique. I wanted the teachers to use precise language to communicate their thinking and reasoning about abstract objects. They were often satisfied to exhibit something they had made and draw empirical conclusions. Should I then tell them what they might have said and how they ought to have reasoned, thereby satisfying my own needs for being in control? Or by doing so would I cover up more than I would uncover? These are real dilemma for all of us in mathematics education.

On Being "Practical"

All of the teachers in the geometry course were highly competent professional educators who were engaged in part-time study (either in the evenings or during summers) while they taught classes during the day. They were serious about their work and genuinely concerned about their students. Thus, they hoped their course work would help them cope more effectively with their difficulties, difficulties that frequently were pressing and severe. Although not all of them expected to find it, they yearned for something of practical value in their mathematics courses. Indeed, they had good reason to believe that their courses would help them in their daily work; this had been a prominent objective when planning and recruiting for the program and also was part of the program evaluation. So there was considerable pressure to deliver something of practical value.

But what kinds of things have 'practical' value? While we—that is the teachers and I—all agreed that the geometry course should make connections to their practices, we did not all have the same understandings of what this meant. At one extreme were the teachers who expected that what they experienced in class should transfer, with only minor modifications, to their classrooms. So, for example, a few teachers said they were unable to use the material they were learning because they did not teach geometry or they only taught young children. Others (like Sam) expressed doubt that their students would be able to handle materials or produce ideas on their own. Still others tried a few things but stopped when they found that they did not "work."

A less extreme position recognized that not everything would transfer directly and therefore some modifications would be necessary. Those in this camp often introduced the same materials as we had done in class, sometimes using the same activities and sometimes different ones. If we built 3-D models in the geometry class, their students were building similar models a few weeks later. If I used tangrams, all of a sudden many of them were using them. If I demonstrated with a computer program, some teachers tried it also. Although this was all to the good, sometimes they tried activities without fully understanding them. Consider the following from José's journal:

> I have never before had the freedom to construct something of my choice. I think my students would (also) enjoy constructing shapes and would welcome this freedom. So this would be a great extension to my lesson. But first I would have to give them notes about the names of shapes and their characteristics.

José seems to have had a powerful experience—albeit one he could have described more graphically—and he would like his students to have a similar experience. But was it necessary for him to introduce the names of concepts before students had experience with these concepts?

A third position sees more practical value in methods and pedagogical techniques than in specific activities. Many teachers, following my lead, began to use groups more frequently, paid attention to the kinds of questions they asked, asked students to keep journals, or planned activities with multiple entry points. Julie (see comments above) provided many good examples, at one point specifically saying that she was trying to adopt my method "of keeping my mouth shut." Those who took this perspective began to reflect on their own teaching. Here are two examples:

> In order for students to have an enriched mathematical experience it needs to be done through discovery. Giving the students everything does not help them. In most cases we are teaching them to wait for the answer because they have learned that in a minute or two we will tell them. It takes away their ability to critically think about a problem and use their skills to determine a rational answer. I am guilty of doing this.
>
> In all the other subjects I teach, I use the text as a springboard, not as the main lesson. Why have I not done the same for mathematics? Why do I teach math differently from other subjects? Something I shall puzzle over during this course to be sure.

There is a fourth position, one that was an important objective of the geometry course. From this perspective, what transfers most powerfully is the emotional experience of learning. Once teachers have experienced math in a new way, they want to recreate this experience for their students. The details of the activity, though important, usually take care of themselves. Numerous quotes from the teachers' journals indicate that many of them came to share this perspective. As one teacher said, "My experiences as a student [make] me to want to create that learning environment in my classroom." Another draws parallels between her experiences in the geometry class and what she would like her students to experience:

> It is really important to find a balance between too much or too little information. I constantly struggle with that in the classroom. I now feel more confident with letting students struggle with their thinking because I have enjoyed being challenged and knowing that answers will require all of my focus. I am trying to [re]create that environment in my classroom.

A third raises important questions about his students:

> Again I wonder if how I am feeling, is very similar to how my students feel. If it is, then what is the appropriate approach that a teacher should take? Should I slow down with material that needs to be covered? Should I call parents to inform them that their children are slacking? Or should I understand them and try to work with them accordingly?

And a fourth adds the idea of trust:

> I am finding that experiencing class like this as a student is going a long way. … I think we need to trust our students more, and let them show that they can do it as well. … With time, experience, and thoughtful reflection, progress should be made, and a new culture of learning can be created.

Because I wanted to reacquaint teachers with who they were as learners of mathematics, I focused on their learning first and then on applications to teaching. As I said in response to one teacher's journal:

> I am glad to see you thinking about how you could adapt the experiences we have in class for use with your students. As you do this, keep in mind that the most important thing is for your students to have the same kind of powerful experiences that you are having in this class. Sometimes you can use what we do directly, but more often than not you may have to make some significant changes. I am not trying to model exactly what you should do with students (though sometimes it might work to copy what I do). Rather, I am trying to change the way you look at math by providing you with powerful experiences that you will then want your students to have. This is why I keep asking you how you felt (for better or worse) about what we are doing in class.

In my view, this builds a stronger connection to practice than focusing on specific things to do tomorrow.

Conclusion

In this chapter, I have argued that reforming mathematics teaching is a complex process that requires a great deal of teachers. Not only must they overcome powerful emotions and well-entrenched habits, they also must change their beliefs about mathematics learning and teaching. To make these changes, it is not sufficient to simply add to their sum total of knowledge. Certainly mathematical content knowledge and pedagogical content knowledge are important, but we also need to attend to teachers' emotional experiences of learning mathematics, for these are what will transfer most powerfully to their classrooms.

At the heart of this view is the notion that teaching is fraught with objective dilemmas and subjective tensions. Teachers have to make judgments, in individual cases, when there are no sure-fire rules or algorithms for making such judgments. Should I allay students' fears and minimize discomfort? Is it time to reduce frustration by giving more clues? Should I slow down or move forward? Cover more or cover up? How we make these judgments depends, in part, on how we negotiate our internal tensions. But this is what keeps teaching stimulating and fresh even if we sometimes make mistakes in the process. Who said teaching was not exciting? It is a daily adventure.

References

Ball, D. L. (1988). *The subject matter preparation of prospective teachers: Challenging the myths*. East Lansing, MI: National Center for Research in Teacher Education.

Ball, D. L., Thames, M. H., & Phelps, G. (2008). Content knowledge for teaching: What makes it special? *Journal of Teacher Education, 59*(5), 389–407.

Cuoco, A., Goldenberg, E. P., & Mark, J. (2010). Contemporary curriculum issues: Organizing a curriculum around mathematical habits of mind. *The Mathematics Teacher, 103*(9), 682–692.

Education Development Center (EDC). (2000). *Connected geometry*. Chicago, IL: Everyday Learning Corporation.

Goldenberg, E. P. (1996). 'Habits of mind' as an organizer for the curriculum. *Journal of Education, 178*(1), 13–34.

Ma, L. (1999). *Knowing and teaching elementary school mathematics: Teachers' understanding of fundamental mathematics in China and the United States*. New York: Lawrence Erlbaum Associates.

NCTM. (2000). *Principles and standards for school mathematics*. Reston, VA: NCTM. (Available at http://www.nctm.org)

NCTM. (2006). *Curriculum focal points*. Reston, VA: NCTM. (Available at http://www.nctm.org)

Pereira, P. (2005). Becoming a teacher of mathematics. *Studying Teacher Education, 1*(1), 69–83.

Schmitt, W. H., et al. (1997). *Many visions, many aims: A cross-national investigation of curricular intentions in school mathematics*. Boston: Kluwer.

Shulman, L. S. (1986). Those who understand: Knowledge growth in teaching. *Educational Researcher, 15*(2), 4–14.

Stigler, J. W., & Hiebert, J. (1999). *The teaching gap: Best ideas from the worlds' teachers for improving education in the classroom*. New York: Free Press.

Chapter 11
Are We Singing from the Same Songbook?

Anne Prescott

When I became a mathematics teacher educator at university, I believed that my previous 20 years of experience as a school teacher and as a supervising teacher of prospective teachers on practicum provided a good basis for helping prospective teachers to become the best teachers they could possibly be. In common with many teachers I saw my role primarily as passing on knowledge about good teaching practices (Zeichner, 2005). I soon realized that being a teacher educator involved much more.

Context

The prospective mathematics teachers at my university have an undergraduate qualification in mathematics and join the program to undertake a 1-year Bachelor of Teaching degree. One group of students has majored in mathematics and have two 3-hour weekly workshops in secondary school mathematics methods over 9 weeks in each of the two semesters. A second group majored in mathematics and science and in each semester half of their workshops are mathematics methods and the other half are science methods. Practicum consists of two 5-week blocks, each in a different school.

Most of my students (80–90%) in the Bachelor of Teaching for Secondary Mathematics are "career-switchers". Richardson and Watt (2006) accurately describe such students as more mature and professional, having developed a well-defined sense of themselves in their previous careers and possessing people skills germane to teaching. Their reasons for becoming teachers at UTS are varied – "teaching is more family oriented", "I want a less stressful job", "I always wanted to be a teacher and now have the opportunity", "I need to get my qualifications recognized in Australia", and so on.

A. Prescott (✉)
Centre for Research in Learning and Change, University of Technology Sydney, Sydney, Australia
e-mail: anne.prescott@uts.edu.au

S. Schuck, P. Pereira (eds.), *What Counts in Teaching Mathematics*, Self Study of Teaching and Teacher Education Practices 11, DOI 10.1007/978-94-007-0461-9_11,

Just as had happened in my school teaching days, I became aware during discussions and from their lesson plans that these students arrived in class with attitudes, understandings, and perceptions of mathematics and mathematics teaching, which were different from the ones I held. As a supervising teacher it had been easy to manage such differences with a single prospective teacher; now, with a class of prospective teachers, these differences between students' understandings and mine created tensions that had to be recognized and addressed. I realized that the prospective teachers and I were marching to a different tune. Tensions existed that needed to be brought to the surface and discussed, not so that the prospective teachers would all be clones, but so that they would realize that teaching is complicated and involves more than simply standing in front of a class telling the class how to do the mathematics.

In this chapter, I discuss the tensions I experience as a mathematics teacher educator concerning the differences in beliefs about mathematics teaching, the differences in ideas the students have about what the university course will provide, and the issues that arise when the prospective teachers are on practicum and then as beginning teachers. Each difference needs to be acknowledged and questioned if the prospective teachers are to gain the most from their course.

The first tension to arise each year is the tension between the prospective teachers' beliefs about mathematics and my beliefs. Because their experiences of mathematics teaching are positive, they believe that the way they were taught is the way all classes should be taught. They often want to be re-taught the subject matter as it is a while since they were in the classroom. This leads to another tension – how should I conduct the classes in order to help them develop as teachers able to deal with any aspect of teaching? When they go on practicum, tensions sometimes develop between the prospective teachers and their supervising teacher as the student tries to implement their newly formed ideas about teaching as advocated in our classes. They sometimes find themselves observing a supervising teacher conducting classes that are more like their past classroom experience and not what has been advocated as good teaching practice at the university. Finally, when they leave the university as beginning teachers, I am aware that their experiences in their schools will not always be positive and the mentoring of beginning teachers is often ad hoc. The tension then is deciding when my contact with these students should cease – does my job finish when they leave their final class? I am not sure that we can ever say that these tensions go away but I believe that we can relieve them by being open about their existence.

Beliefs About Mathematics Teaching

The Tension

Long before they begin their teacher education course, prospective teachers have a well-developed view about mathematics, about teaching and learning mathematics, and about schools because they have spent years in mathematics classes as students themselves. These beliefs are naturally grounded in their own classroom experience where often a rule was taught and then numerous examples of using the rule were practised. These ideas were confirmed when they attended mathematics lectures at university. The experiences they have had with mathematics shape their feelings about the subject and about themselves in relation to it (Ball, 1988). There is much evidence in the literature in mathematics education that teachers see mathematics as a fixed and sequential body of knowledge that is most effectively learned by rote, and that consists of algorithmic and repetitive procedures (Crawford, Gordon, Nicholas, & Prosser, 1994; Nyaumwe, 2004; Schuck & Foley, 1999). There is a consequent emphasis on instrumental learning rather than relational learning (Skemp, 1976) resulting in:

- mathematics curricula driven by computational skill as the major goal;
- mathematical knowledge as rule bound and unconnected;
- teaching as telling and learning as memorizing (Even & Lappan, 1994).

Therefore, in many mathematics classrooms, there is a reliance on a "shallow teaching syndrome" (Stacey, 2003; Vincent & Stacey, 2008) where students complete a large number of repetitive, low complexity problems, often by blindly copying procedures at the direction of their teachers. Stein and Lane (1996) note that there is a tendency for teachers "to proceduralize tasks due to time constraints" and to "perform the most demanding parts of tasks for students" (p. 60).

Ball (1990) found that many teachers also believe that mathematics is a collection of discrete bits of procedural knowledge that are rarely connected and therefore require innate ability to master. Furthermore, there is a close relationship between beliefs about mathematics and classroom practice (Nisbet & Warren, 2000), and teachers' beliefs are often strongly held and notoriously difficult to change (Wilson & Cooney, 2000). Rather than describe learning as a process of passively receiving new knowledge, sociocultural theorists understand learning to take place through activity by individuals and groups in particular tasks. For example, Schön (1983) described the acquisition of professional knowledge as "knowing-in-action". Therefore, it may be difficult for teachers to adopt practices in the classroom, such as a problem-solving approach where students construct mathematical ideas for themselves, that depart significantly from those which they experienced as learners (Szydlik, Szydlik, & Benson, 2003). The conflict arises when learning is seen, not as a purely theoretical exercise but rather as participation in specific activities to gain new insights and skills (Adler, 1998).

Most, if not all, prospective mathematics teachers enjoyed mathematics at school, where they responded favourably to traditional forms of teaching in their own education. Moreover, they tend to believe that their own students will react just as positively to a similar direct instruction model. Thus they find it difficult to imagine a need to teach in any other way (Ball, 1988). Many classroom students respond to this style of teaching with boredom and discouragement, developing a belief that mathematics depends on some innate ability and believing it will never be useful once they leave school.

Notions of what constitutes "good teaching" are thus formed early in a person's schooling and can prove difficult to shake, particularly because they are usually based on school experiences and the personalities of individual teachers rather than on pedagogical principles (Lortie, 1975). Even prospective teachers' initial observations are necessarily from the students' perspective, so the meanings that are attached to them lack any real appreciation of the complexities of teaching.

Pre-service secondary mathematics teachers often believe that a good knowledge of their subject is all that is necessary to be a good teacher (Brookhart & Freeman, 1992) and that the mathematics is easy (Ball, 1990). However, when asked to explain concepts, they also admit that they "do it by rote", memorizing rules and procedures to the point where mathematical explanations become restatements of the rule (Ball, 1990; Foss & Kleinsasser, 1996).

On the other hand, I believe that the contexts in which learning takes place are critical because they help to shape the learning that occurs within them (Franke & Kazemi, 2001). Social, historical, cultural and physical settings play an integral part in what is learned and how it is learned. In many cultures, learning is a social activity which happens in communities of practice (Wenger, 1998) where people engage in some collective activity or shared enterprise. Learning is said to occur through increasing participation in joint activities aligned to common goals, purposes, means and ends (Lave & Wenger, 1991).

I see the students' view of mathematics as limiting their ability to teach effectively. Their view of teaching and mathematics is situated in a transmissive approach to teaching which usually involves telling the students the rule and then working in silence as the students practise applying the rule. I do not believe that the student teachers understand that although this approach may have worked for those who were good at mathematics, it does not help students who are less able at mathematics. Students who thrived in school therefore continued learning in this way when studying mathematics at university, while less able students floundered and lost interest in mathematics. Given that prospective teachers are likely to be teaching many students in this latter category, it is important that we disrupt this belief throughout the course.

Relieving the Tension

Changing teachers' beliefs about mathematics is difficult. It may not be until student teachers experience their first practicum that they realize that the lessons in which they just taught rules and the students completed examples were not their better lessons. The difficulties are evident in a study of prospective teachers' views of teaching conducted by Cavanagh and myself (Prescott & Cavanagh, 2006). One student teacher described a good lesson as one which would "have lots of student activities", engaging the students in practical activities to try to link the concepts to the "real world" and contrasted this with unsatisfactory lessons where "the teacher is sitting out the front assigning exercises and just letting the students do the exercises and maybe doing a few examples on the board without really engaging the students". However, when asked how he might teach a lesson on decimals, he responded:

> I'd start with the concepts, I guess. So I'd write up a heading on the board say, for example, adding decimals and I'd write the theory behind it. You know, you need to put the point above the point and then I'd run through some examples. But I would make sure the examples would have a real-life context to it like money examples and money problems. And then I would just continue on with the other theory.

The tension in effecting change in his attitudes is evident – he recognizes the good lesson, but his limited personal experience of a more student-centred approach is insufficient to ensure that he examines his deeply rooted views about teaching and learning mathematics.

Such beliefs about the nature of mathematics teaching are known to be difficult to shift (Brown & Borko, 1992; Cooney et al., 1998) and most research has shown that these beliefs seldom change during teacher education courses (Foss & Kleinsasser, 1996; Kagan, 1992). When a prospective teacher's beliefs are at odds with the underlying philosophy of the course, it is unlikely that any change will occur (Bright & Vacc, 1994) and such a situation may even result in initial biases becoming stronger (Szydlik et al., 2003).

Posner, Strike, Hewson and Gertzog (1982) suggested that for existing beliefs to be replaced, the new belief must be intelligible and appear plausible. If prospective teachers cannot see a reason to change, they often alter the new idea to fit their original long-held beliefs. Ball (1990) concluded that these beliefs not only shape how prospective teachers teach but also how they approach learning to teach. Therefore, it seems essential that pre-service courses challenge prospective teachers' views of mathematics so they become open to the more collaborative, creative discipline that is mathematics (Even & Lappan, 1994). The tension between what the prospective teachers and I believe about mathematics and mathematics teaching must be confronted.

Like most teacher educators I want to assist my students to develop a different style of teaching from the traditional practices that they themselves most often experienced and observed. Telling students about the range of variation in ideas about learning mathematics will not, of itself, broaden their conceptions, but it will make it easier for students to think more deeply about their own views of learning (Reid, Wood, Smith, & Petocz, 2005). Therefore, I model non-traditional practices allowing the students to experience a different style of teaching (McGinnis, Watanabe, & Roth McDuffie, 2005). I discuss progression from the concrete to the abstract, and we discuss different learning styles. In the less academic classes, especially, it is important that activities with concrete objects form the first step in the classroom to enable the students to understand the connections between the mathematics of the classroom and the mathematics in their daily lives. I seek to show the students that effective mathematics teachers make their classes interesting through problems and activities that are related to the "real world" of their students, and I encourage students to explore, investigate, and construct solutions on their own. I use a number of instructional strategies including concrete materials, technology and cooperative group activities (McGinnis, Watanabe, & Roth McDuffie, 2005). Mathematical games, puzzles and stories are useful to enable children to make these connections and to build upon their everyday understandings.

We often talk about the less able students in schools – usually these prospective teachers have been successful mathematical learners and have little experience of students who struggle with mathematics. They cannot envision the possibility that their students might develop maths anxiety because they have always enjoyed mathematics. So we look at how it might develop in students and importantly, we also look at how we might help students overcome it or at least deal with it in a positive way. In this, I often use examples from my teaching because such stories about teaching practice are a beneficial source of information about teaching and also a worthwhile reflective exercise (McMahon Giles & Moore, 2006).

I want prospective teachers to be more open to learning mathematics in groups so they see it as a natural part of learning mathematics. Much of their experience of mathematics is as an individual getting the answer – the answer is everything. So we begin in the first workshop with a series of simple Year 7 problems where algebra is not needed, indeed it is forbidden. This forces students to think of alternatives. The students work in pairs to solve one of the problems and then they have to explain their solution to the rest of the class. Many have never discussed solutions of mathematical problems, and, as I have only handed one set of the questions to each pair of students, they find themselves in a situation where they must talk to each other and agree on a solution. I spend much of the workshop encouraging them to work together rather than working as individuals in silence, and it becomes part of the humour of the class when I admonish them for being too quiet!

It takes some time for the prospective teachers to recognize the difference between instrumental and relational learning (Skemp, 1976) but we look for examples in the work we do in class, especially when we discuss the Working Mathematically strand of the NSW Years 7–10 Mathematics syllabus (NSW Board of Studies, 2003). This strand promotes the use of problem-solving and mathematical investigations through five interrelated student processes: questioning, applying strategies, communicating, reasoning and reflecting. Teachers are encouraged to cover the syllabus content by developing and using a variety of activities that allow students to engage with the five processes. The syllabus is organized so that each learning outcome has a working mathematically outcome. When the students write lesson plans, they are expected to include these outcomes as well as previous knowledge required. In this way, they are seeing that there are connections within and between mathematics topics.

Depending on their previous careers, we often have the benefit of their experience in these careers when they write their lesson plans. One former stockbroker wrote a lesson plan on shares for a General Mathematics lesson, and the 20-minute excerpt of the lesson that she presented to the class was outstanding. The students who have IT experience raise the standard of the lesson presentations, with one using Geogebra to explain the circle theorems. Later, when he was on practicum, he ran a workshop for the mathematics teachers in the school on using Geogebra, including installing it on all the computers in the school.

Mathematics Methodology Class Content

The Tension

Beginning with Shulman in 1986, many researchers have categorized teacher knowledge into four areas – general pedagogical knowledge, subject matter knowledge, mathematical pedagogical content knowledge (MPCK) and knowledge of context (e.g. Ball, Hill, & Bass, 2005; Goulding & Suggate, 2001; Kahan, Cooper and Bethea, 2003; Lim-Teo, Chua, Chenag, & Yeo, 2007; Mewborn, 2001; Stacey et al., 2001). However, students believe that subject matter knowledge is all important, particularly at the beginning of their course. They believe that other subjects in the Bachelor of Teaching program cover general pedagogical knowledge and knowledge of context leaving the mathematics methodology subjects to concentrate on subject matter knowledge and MPCK.

Initially, prospective teachers believe that we should be concentrating on subject matter knowledge and, of course, this is directly linked to their view of education and of mathematics. This tension arises from many of the prospective teachers being career switchers and so it is some time since they studied mathematics. The issue is highlighted on practicum when they come across mathematical concepts they have to learn before they can teach them, or their supervising teachers find it difficult to understand how a person with a mathematics degree, albeit from some years ago, can make mistakes with basic mathematics.

Early in their course, the prospective teachers begin to realize that MPCK is a complex concept integrating generic pedagogical knowledge, mathematics teaching methodology as well as knowledge of the discipline of mathematics (Lim-Teo, Chua, Cheang, & Yeo, 2007) and thus MPCK determines the teachers' choice of examples, explanations, exercises, items and reactions to children's work (Ball, 2000).

An important consequence for teaching styles, in universities and in schools, is the need to focus more on the *process* than on the *product* (Johnston, 2003). Although it would be easy to focus on the product instead of focusing on the process, I want my students to develop a deep understanding of MPCK. Unless they are involved in constructive experiences that focus on processes, students will simply learn a random association of arbitrary rules (Gould, 2004) concentrating on the product.

Relieving the Tension

Because many of the prospective teachers are changing careers, they are often concerned that they have forgotten the mathematics they will need. Taking a basic

Year 7 arithmetic test without a calculator confirms their suspicions. While we do not have time to cover all the syllabus content (from one Year 7–10 syllabus and four Year 11–12 syllabuses), I have found that we can cover much of the content by judiciously choosing topics for the prospective teachers to present as lessons and by looking in detail at topics that school students traditionally find difficult or which would be new to the prospective teachers. For example, we look at all the finance sections in the senior syllabuses, paying particular attention to the difficult Financial Mathematics section of the General Mathematics syllabus, which is new to most of the prospective teachers.

When the student teachers prepare a lesson plan, their classmates comment on it, and so do I, and then they rework the lesson plan to be taught in class. This final lesson plan and their class presentation are the assessment task. The lesson plans are submitted online as a resource for when they are teaching. This system also allows me to discuss with the class where the misconceptions may arise and where school students often make mistakes. The comments by their peers are designed to prepare the student teachers for times when they will be busy teachers downloading lesson plans from the Internet. The prospective teachers are getting practice in assessing whether a lesson plan is appropriate for the assigned class. The last process in the lesson plan assessment is a reflection by the student about the lesson and its presentation. Having taught 20 min of their lesson, they begin to realize that even with a very kind class of their peers, too much talk and too little activity is problematic. As a result, I often see them confronting the tension between their beliefs and their teaching (Lomas, 2009).

As mentioned earlier, maths anxiety is often a problem in less academic classes. Many prospective teachers do not understand that anxiety, let alone the phobia that may develop. When a mathematics teacher emphasizes procedure and knowledge of formulas over understanding, students often become anxious and a silent (and sometimes, not so silent) majority gives up, resigned to failing maths.

I strive to convince my students that understanding when and how a formula is to be used is always more important than recalling the technique from memory. I believe an important goal of mathematics teaching should be creating a taste for mathematics so it is used and enjoyed. We emphasize problem solving and activities that engage students and offer a sense of success.

On Practicum

The Tension

It is particularly important for learners to encounter consistent messages and theories that can help them make sense of the phenomena they experience and observe. Mixed messages and contradictory theories serve only to confuse. (Hammerness, 2006). The practicum is set up to allow students to practise skills and modes of thinking initiated at the university. Still, an oft-repeated comment by supervising teachers is that prospective teachers should forget all the rubbish they learn at university because it is in the classroom that one learns to be a teacher (McMahon Giles & Moore, 2006). While we know that there should be coherence between the learning at university and the practicum, often there is not.

Prospective teachers frequently characterize their supervising teachers as "traditional" and claim that it is difficult to experiment with teaching ideas because the supervising teacher is dismissive of such an approach. Sometimes this takes the form of the supervising teacher insisting that a plan incorporating group work or problem-based activities be changed to a more teacher-centred mode of delivery, reasoning that such teaching methods do not allow for the completion of a sufficient number of practice exercises during lessons (Cavanagh & Prescott, 2007). Lortie (1975) calls this the "apprenticeship of observation" and believes it to be a major influence in shaping teachers' conceptions of the teaching role. Formal teacher education is viewed as having little impact in altering the cumulative effects of such socialization.

Often, the prospective teacher is aware of the tension between theory and practice and, lacking any authority to do otherwise, usually complies with classroom practices (Waghorn & Stevens, 1996). They teach in the preferred style of their supervising teacher even though they do not agree or even like what they are doing. They set aside their own preferred methods and ideas, suspend their own judgements and adopt what they often think of as less desirable ways of teaching. The pressure to conform to the supervising teacher's style can also be a factor in determining the kind of final practicum report that each prospective teacher receives. The report is a high-stakes document in the minds of the prospective teachers because they use it in job interviews as evidence of their teaching capabilities. The prospective teachers often decide that the best way to guarantee a good report is to follow closely the supervising teacher's advice which usually means teaching in a traditional way (Cavanagh & Prescott, 2007).

Prospective teachers often interpret their practicum experience in fairly simplistic or idealistic terms, conceiving teaching – especially classroom management – as primarily technical competence rather than as a process of on-going decision-making focused on student learning (Cavanagh & Prescott, 2007). This becomes evident when I ask prospective teachers about the lesson they have just presented to a class. They rarely discuss the learning by the students and usually worry about their own performance.

Many of the prospective teachers want to try alternate teaching strategies, such as group work, only to find that their students, who may have never worked in groups, react against any change from the traditional classroom routines to which they have become accustomed. Unfortunately, the prospective teachers' first, tentative steps in using alternative teaching strategies are not always successful and often result in minimal student participation or learning. Once their initial attempts at using alternative teaching strategies fall short of their expectations, they are often reluctant to try them again, particularly when they perceive that the supervising teacher, who later writes their final practicum report, is also unimpressed by these lessons (Cavanagh & Prescott, 2007).

Relieving the Tension

This tension came to a head one practicum when a couple of the prospective teachers demanded to be allowed to try new ideas, while their supervising teachers wanted to inculcate the pre-service teacher into the traditional norms of the status quo (Cavanagh & Prescott, 2007). Given the supervising teacher writes the report at the end of practicum, a report that is used to gain employment, we had a tense time on the practicum as I negotiated with the supervising teacher on behalf of the prospective teachers. After experiencing this stressful practicum, I now ensure this sort of situation is discussed with prospective teachers before they go on practicum – we discuss ways that we can undertake less traditional modes of teaching like group work and technology lessons when the supervising teachers are less than supportive. I suggest that the prospective teachers take it slowly. Students are often sitting in pairs in the classroom so I suggest that an activity where the pairs look together at a particular activity is a way of introducing a different style of learning and teaching. I spend a lot of time in our workshops modeling ways of working in the classroom that will be acceptable to a supervising teacher who has a more traditional style of teaching. We also acknowledge that all teachers are different just as all students are different and that it is important for the prospective teachers to observe many classes and not just their supervising teacher's classes to gain ideas and teaching strategies other than one teacher's perspective.

On the other hand, it has been reported that much of what was learned in teacher education courses is "washed out" (Zeichner & Tabachnick, 1981) once teachers enter their own classrooms. The prospective teachers need to be comfortable in the teaching strategies we advocate at the university so that when they are in control of their own classroom, they are confident in trying new ideas. Writing and presenting lessons, participating in the discussions about pedagogy, teaching on practicum, and observing many teachers' classes enable prospective teachers to develop their own style and the confidence to try alternative teaching strategies.

On Becoming Teachers

The Tension

The last issue I have to consider is when do I let go of my students? It is well known that teaching is a difficult and demanding task and that a beginning teacher's first year is stressful. It is not surprising that the first year of teaching can be an important determinant in a teacher's career and that "surviving" the beginning years of teaching has a major impact on continuing in the profession (Loughran, Brown, & Doecke, 2001). The first year is often a time of disillusionment, failure and shattered idealism. Beginning teachers find themselves in survival mode as they cope with the daily realities of the classroom, often with a teaching load that experienced teachers would find challenging.

The tension between looking for support and being regarded as an effective teacher can be problematic for the beginning teachers. Their perceptions of themselves as teachers are also coloured by the feeling that they never have a spare moment. The administrative details of their job and activities such as playground duty take them away from their work as teachers so even if they are keen to produce excellent lessons, they spend so much time on details outside the classroom that the idea of being a reflective teacher usually does not happen (Prescott & Cavanagh, 2008).

Criticisms of the teacher education program are often heard because of the expectation that graduates from a teacher education program emerge as full-blown experts. Beginners are often expected to perform as well as 30-year veterans, sometimes with more difficult classes than their more experienced colleagues (Russell, McPherson, & Martin, 2001). It almost seems to be a sport to give any new person all the classes that the other teachers do not want. It takes time for any new teacher in a school to become established – when this new teacher is a beginning teacher, it is no wonder that they often struggle. On many occasions my ex-students discuss how their classes include three or four (out of five) less academic classes with major classroom management issues. Add to this their desire to be seen as a competent teacher and their reticence in asking for help, and it becomes clear why the attrition rate amongst new teachers is high.

Often these new teachers do not have many support structures to make the transition into full-time teaching less demanding and difficult. There is a mythical notion in teaching that "ordeal by fire" helps to sort out those who can and those who cannot teach, but this notion blurs the difference between coping with support during professional transition and an individual's "ability" (Loughran, Brown, & Doecke, 2001). Some schools establish good mentoring systems that work well, but others do not. One of my ex-students had no mentoring within the school, but she had a mentor assigned from outside the school. When she asked for some ideas in handling a particularly difficult class in this particularly difficult school, she was told that this mentor did not have time. At this point she felt abandoned. Still, we discussed some options for managing the class. These proved successful, and I am pleased to say she is still teaching at that school four years later.

So what happens when my students become beginning teachers and I hear they are floundering, often with little help in the school?

Relieving the Tension

After my first year as a mathematics teacher educator, I found out that one of my promising prospective teachers lasted two weeks in a school. I was disappointed that when he was having difficulties in the classroom he did not think to contact me for help. From that time I set out to maintain contact with each group for at least their first year of teaching.

The influence of the cohort group cannot be underestimated, particularly because it facilitates an environment in which students begin to feel part of a strong professional culture (Graber, 1996). By getting the prospective teachers working in groups for assignments and having them critique each other's lessons, I have tried to establish the group as mentors to each other, a position that goes further than just being friends or working on assignments together. It is particularly noticeable in the second semester of their course, when each week there is a set of mathematical problems to be solved. Not only do the groups work on the problems together but they also make sure that each person in the group understands the work. Students are chosen at random to present the solution as if to a class in a school so they work together to make sure everyone is "up to speed."

This environment of collaboration extends beyond the university to practicum support. Most of the students are the only student teacher in their school so they continue to support each other after school each day. They share ideas for teaching and classroom management, and reassure each other about how difficult teaching is.

As an example, two teachers from the 2008 cohort were a great support to each other during their first year of teaching. One was in a school known for its classroom management difficulties, but he had support from the school administration. The other was in a school with fewer behaviour problems, but she felt she had no support from the administration. Every afternoon they each had a long drive home where they debriefed each other over their hands-free mobile phones. They believed that the only thing that kept them in teaching was this regular discussion about the school day as they drove home.

Within the first month of first term, we meet over coffee and cake for a couple of hours at the university, to find out how everyone is going. Because they are used to mentoring each other, they are honest with each other about the situations they are in and they are all open to suggestions. I rarely need to offer advice because they all help each other. Those without jobs in schools also come, and we have found that many go away with some casual teaching opportunities if not the possibility of full time work, because vacancies are also discussed.

More often than not, the discussion comes around to classroom management with sharing of ideas that worked and did not work. Again, I rarely offer advice. They have similar experiences of a preponderance of less able classes and full teaching loads so all coping mechanisms are freely offered and gratefully received. Possible issues and concerns are highlighted during the teacher education course, but they are difficult to address then because we cannot duplicate the reality of the actual world of beginning teaching (Loughran, Brown, & Doecke, 2001). It is only with the experience of the classroom that they understand the issues.

In 2009 one of these new teachers offered advice about accomplishing accreditation with the NSW Institute of Teachers. In his school, he had been mentored well and not had difficulties, so his advice was well received by those not receiving help. This sharing between the cohort has become a natural part of the teacher education program to the point they are surprised when they do not find it in the schools in which they teach.

We meet in all four terms of the school year and the 2009 cohort celebrated their first year of teaching with a barbecue at the university. It was a wonderful celebration with partners and children also attending. They were so proud of their achievements. It was also important for us all to recognize that they had completed a year of teaching and to hear that the next year was going to be better because of what they had learned during that first year.

Conclusion

When I first started as a mathematics teacher educator, I believed that my experience was the most important element in my ability to undertake the job. It did not take long to see that a number of tensions existed between how I approached the classes and how my students approached the classes. Preparing them for teaching was far more complicated than I imagined, but my awareness of the tensions that arose helped me to reflect on my practice and consider ways of enhancing it. These tensions are certainly not resolved – I find each year that a new aspect of each tension is evident – but at least I am looking to see if we are singing from the same songbook, and then from the same page. Each cohort is different, depending on the personalities and backgrounds of the participants, but then, as a teacher, I have always been aware of differences between classes and between students and the need to manage these differences. The tensions feel bigger than just individual differences – they are at the core of what we are about as educators.

Our beliefs about mathematics determine how we approach teaching and, in my case, determine how I approach the workshops with the prospective teachers. The tension that arises in class could make the students feel I am putting down their teachers when I try to broaden their ideas about what mathematics is and what teaching is all about. Indeed, that is exactly what happened when I started my first education lecture in my undergraduate degree – the lecturer told us all we had had a very poor education. Now, for students who thought their education and their teachers were wonderful, this was not welcome news and it was only later, when I had my own classes, that I realized what the lecturer had meant. This is particularly important when I prepare the prospective teachers for practicum. When the supervising teacher on practicum is going to write a report on the prospective teacher, it is important that they form a working relationship quickly because the report is used to gain employment. We, therefore, spend time discussing these issues. I have to admit that I am not sure of the answers. How should they react if their supervising teacher is a very traditional teacher? How can they make the most of the experience? How can they use some of the ideas that we have developed in university classes when the supervising teacher is not supportive?

There are not many professions that expect a brand new recruit to handle the job in the same way as a 30-year veteran, but the teaching profession has that expectation. This places enormous pressures on beginning teachers and has implications for teacher educators. How do we prepare our prospective teachers for the rigours of their first year of teaching? How can we help them stay in the profession if they have little support in the school? I cannot change what is happening in the school so my solution has been to keep in contact. Is it helpful? The fact the beginning teachers keep coming back for support each meeting suggests it is worthwhile – after all the coffee and cake is not that good! Can I keep in contact over many years? Some beginning teachers still contact me for support even some years later, and I enjoy seeing them at conferences. I am not sure that I have the solutions, but I believe that acknowledging the tensions is a start. It truly is a complicated process.

References

Adler, J. (1998). Lights and limits: Recontextualising Lave and Wenger to theorise knowledge of teaching and of learning school mathematics. In A. Watson (Ed.), *Situated cognition and the learning of mathematics* (pp. 161–177). Oxford: Centre for Mathematics Education Research.

Ball, D. L. (1988). Unlearning to teach mathematics. *For the Learning of Mathematics*, *8*(1), 40–48.

Ball, D. L. (1990). The mathematical understandings that prospective teachers bring to teacher education. *The Elementary School Journal*, *90*(4), 449–466.

Ball, D. L. (2000). Bridging practices: Intertwining content and pedagogy in teaching and learning how to teach. *Journal of Teacher Education*, *51*, 214–247.

Ball, D. L., Hill, H. C., & Bass, H. (2005). Knowing mathematics for teaching. *American Educator*, *Fall*, 14–17.

Bright, G. W., & Vacc, N. N. (1994). *Changes in undergraduate preservice teachers' beliefs during an elementary teacher certification program*. New Orleans, LA: Paper presented at the annual meeting of the American Educational Research Association.

Brookhart, S. M., & Freeman, D. J. (1992). Characteristics of entering teacher candidates. *Review of Educational Research*, *62*(1), 37–60.

Brown, C., & Borko, H. (1992). Becoming a mathematics teacher. In D. A. Grouws (Ed.), *Handbook of research on mathematics teaching and learning* (pp. 209–239). New York: Macmillan.

Cavanagh, M., & Prescott, A. (2007). Professional experience in learning to teach secondary mathematics: Incorporating pre-service teachers into a community of practice. In J. Watson & K. Beswick (Eds.), *Mathematics: Essential research, essential practice* (Vol. 1, pp. 182–191). Proceedings of the 30th annual conference of the Mathematics Education Research Group of Australasia. Adelaide, SA: MERGA.

Cooney, T. J., Shealy, B. E., & Arvold, B. (1998). Conceptualizing belief structures of preservice secondary mathematics teachers. *Journal for Research in Mathematics Education*, *29*(3), 306–333.

Crawford, K., Gordon, S., Nicholas, J., & Prosser, M. (1994). Conceptions of mathematics and how it is learned: The perspectives of students entering university. *Learning and Instruction*, *4*, 331–345.

Graber, K. (1996). Influencing student beliefs: The design of a "high-impact" teacher education program. *Teaching and Teacher Education*, *12*, 451–466.

Even, R., & Lappan, G. (1994). Constructing meaningful understanding of mathematics content. In D. B. Aichele & A. F. Coxford (Eds.), *Professional development for teachers of mathematics*. Reston, VA: NCTM.

Foss, D. H., & Kleinsasser, R. C. (1996). Preservice elementary teachers' views of pedagogical and mathematical content knowledge. *Teaching and Teacher Education*, *12*(4), 429–442.

Franke, M. L., & Kazemi, E. (2001). Teaching as learning within a community of practice: Characterizing generative growth. In T. Wood, B. Scott Nelson, & J. Warfield (Eds.), *Beyond classical pedagogy: Teaching elementary school mathematics: The nature of facilitative change* (pp. 47–74). Mahwah, NJ: Lawrence Erlbaum.

Gould, P. (2004). The language of working mathematically. *Scan, 23*(1), 4–7.

Goulding, M., & Suggate, J. (2001). Opening a can of worms: Investigating primary teacher's subject knowledge in mathematics. *Mathematics Education Review, 13*(March), 41–54.

Graber, K. (1996). Influencing student beliefs: The design of a "high-impact" teacher education program. *Teaching and Teacher Education, 12*, 451–466.

Hammerness, K. (2006). From coherence in theory to coherence in practice. *Teachers College Record, 108*(7), 1241–1265.

Jaworski, B., & Gellert, U. (2003). Educating new mathematics teachers: Integrating theory and practice, and the roles of practising teachers. In A. Bishop, M. Clements, C. Keitel, J. Kilpatrick, & F. Leung (Eds.), *Second international handbook of mathematics education* (pp. 823–876). Dordrecht, the Netherlands: Kluwer Academic Publishers.

Johnston, B. (2003). From catechism to critical literacy: Figuring out a challenging numeracy. *Fine Print, 26*(2), 9–15.

Kagan, D. M. (1992). Professional growth among preservice and beginning teachers. *Review of Educational Research, 62*(2), 129–169.

Kahan, J. A., Cooper, D. A., & Bethea, K. A. (2003). The role of mathematics teachers' content knowledge in their teaching: A framework for research applied to a study of student teachers. *Journal of Mathematics Teacher Education, 6*, 223–252.

Lave, J., & Wenger, E. (1991). *Situated learning: Legitimate peripheral participation*. Cambridge, MA: Cambridge University Press.

Lim-Teo, S. K., Chua, K. G., Cheang, W. K., & Yeo, J. K. (2007). The development of diploma in education student teachers' mathematics pedagogical content knowledge. *International Journal of Science and Mathematics Education, 5*, 237–261.

Lomas, G. (2009). Pre-service primary teacher perceptions of mathematics lecturers' practice: Identifying issues for curriculum development. *Mathematics Teacher Education and Development, 11*, 4–21.

Lortie, D. C. (1975). *Schoolteacher: A sociological study*. Chicago, IL: University of Chicago Press.

Loughran, J., Brown, J., & Doecke, B. (2001). Continuities and discontinuities: The transition from pre-service to first-year teaching. *Teachers and Teaching: Theory and Practice*, *7*(1), 7–23.

McGinnis, J. R., Watanabe, T., & McDuffie, R. (2005). University mathematics and science faculty modeling their understanding of reform based instruction in a teacher preparation program: Voices of faculty and teacher candidates. *International Journal of Science and Mathematics Education, 3*(3), 407–428.

McMahon Giles, R., & Moore, A. L. (2006). Teacher educators return to the classroom. *Teachers College Record*, August 17, 2006. ID Number: 12673.

Mewborn, D. (2001). Teachers content knowledge, teacher education, and their effects on the preparation of elementary teachers in the United States. *Mathematics Education Research Journal, 3*, 28–36.

NSW Board of Studies. (2003). *Years 7–10 Mathematics Syllabus NSW*. Sydney: NSW Board of Studies.

Nisbet, S., & Warren, E. (2000). Primary schools teachers' beliefs relating to mathematics, teaching and assessing mathematics and factors that influence those beliefs. *Mathematics Teacher Education and Development, 2*, 34–47.

Nyaumwe, L. (2004). The impact of full time student teaching on preservice teachers' conceptions of mathematics teaching and learning. *Mathematics Teacher Education and Development, 6*, 23–36.

Posner, G. J., Strike, K. A., Hewson, P. W., & Gertzog, W. A. (1982). Accommodation of a scientific conception: Toward a theory of conceptual change. *Cognition and Instruction, 66*, 211–227.

Prescott, A., & Cavanagh, M. (2006). An investigation of pre-service secondary mathematics teachers' beliefs as they begin their teacher training. In P. Grootenboer, R. Zevenbergen, & M. Chinnappan (Eds.), *Identities, cultures and learning spaces* (Vol. 2, pp. 424–431). Proceedings of the 29th annual conference of the Mathematics Education Research Group of Australasia. Adelaide, SA: MERGA.

Prescott, A., & Cavanagh, M. (2008). *A situated perspective on learning to teach mathematics*. Paper presented at the Proceedings of the 31st annual conference of the Mathematics Education Research Group of Australasia, Brisbane.

Reid, A., Wood, L. N., Smith, G. H., & Petocz, P. (2005). Intention, approach and outcome: University Mathematics students' conceptions of learning Mathematics. *International Journal of Science and mathematics Education, 3*, 567–586.

Richardson, P. W., & Watt, H. M. G. (2006). Who chooses teaching and why? Profiling characteristics and motivations across three Australian universities. *Asia-Pacific Journal of Teacher Education, 34*(1), 27–56.

Russell, T., McPherson, S., & Martin, A. K. (2001). Coherence and collaboration in teacher education reform. *Canadian Journal of Education, 26*(1), 37–55.

Schön, D. (1983). *The reflective practitioner: How professionals think in action*. New York: Basic Books.

Schuck, S., & Foley, G. (1999). Viewing mathematics in new ways: Can electronic learning communities assist? *Mathematics Teacher Education and Development, 1*, 22–37.

Shulman, L. S. (1986). Those who understand: Knowledge growth in teaching. *Educational Researcher, 15*, 4–14.

Skemp, R. R. (1976). Relational understanding and instrumental understanding. *Mathematics Teaching, 77*, 20–26.

Stacey, K. (2003). The need to increase attention to mathematical reasoning. In H. Hollingsworth, J. Lokan, & B. McCrae (Eds.), *Teaching mathematics in Australia: Results from the TIMSS 1999 Video Study*. Melbourne. VIC: Australian Council for Educational Research.

Stacey, K., Helme, S., Steinle, V., Baturo, A., Irwin, K., & Bana, J. (2001). Preservice teachers' knowledge of difficulties in decimal numeration. *Journal of Mathematics Teacher Education, 4*, 205–225.

Stein, M. K., & Lane, S. (1996). Instructional tasks and the development of student capacity to think and reason: An analysis of the relationship between teaching and learning in a reform mathematics project. *Educational Research and Evaluation, 2*(1), 50–80.

Szydlik, J. E., Szydlik, S. D., & Benson, S. R. (2003). Exploring changes in pre-service elementary teachers' mathematical beliefs. *Journal of Mathematics Teacher Education, 6*, 253–279.

Vincent, J., & Stacey, K. (2008). Do mathematics textbooks cultivate shallow teaching? Applying the TIMSS video study criteria to Australian eighth-grade mathematics textbooks. *Mathematics Education Research Journal, 20*(1), 82–107.

Waghorn, A., & Stevens, K. (1996). Communication between theory and practice: How student teachers develop theories of teaching. *Australian Journal of Teacher Education, 21*(2), 70–80.

Wenger, E. (1998). *Communities of practice: Learning, meaning and identity*. Cambridge, MA: Cambridge University Press.

Wilson, M., & Cooney, T. J. (2000). Mathematics teacher change and development. The role of beliefs. In G. Leder, E. Pehkonen, & G. Törner (Eds.), *Beliefs: A hidden variable in mathematical education?* (pp. 127–148). Dordrecht: Kluwer Academic Publishers.

Zeichner, K. (2005). Becoming a teacher educator: A personal perspective. *Teaching and Teacher Education, 21*, 117–124.

Zeichner, K., & Tabachnick, B. (1981). Are the effects of university teacher education 'washed out' by school experience? *Journal of Teacher Education, 32*(3), 7–11.

Chapter 12
Adding Value to Self and Content in Mathematics Education: Working in a Third Space

Peter Pereira and Sandy Schuck

> *[Mathematics teaching] is very complicated, subtle, and takes a lot of time. It is easy to be an elementary school teacher, but it is difficult to be a good elementary school teacher.*
>
> (Ma, 1999, p. 136)

The Complexity of Maths Teaching

All of the chapters in this book illustrate complexities that face mathematics educators on a daily basis. Whether they work with students of maths, elementary or secondary teachers, in school or university settings, with another or by themselves, maths educators face a difficult task that is full of pitfalls and possibilities. By sharing the complexities we have encountered, we have sought to include you, the reader, in the dialogue between self and content in which we, as mathematics educators, regularly engage. It is our hope that by doing so, we will encourage you to join the dialogue.

In contrast to our approach, conventional wisdom assumes that mathematics teaching is a straightforward and easily done activity. The so-called deficiencies in maths teaching are obvious to all outside the profession; they are thought to be due to a lack of content knowledge or an inability to teach. The remedies are clear: to address the first deficiency, mathematics teachers only need to learn more content, something they clearly lack if anecdotes and measures of teachers' knowledge of mathematics are to be believed. To address the second deficiency, mathematics teachers need more pedagogy, specifically focused on mathematics instruction, especially pedagogy that will increase test scores. Another commonplace 'wisdom' is that any college graduate in mathematics can learn to be a mathematics teacher

P. Pereira (✉)
School of Education, DePaul University, Chicago, IL, USA
e-mail: ppereira@condor.depaul.edu

S. Schuck, P. Pereira (eds.), *What Counts in Teaching Mathematics*, Self Study of Teaching and Teacher Education Practices 11, DOI 10.1007/978-94-007-0461-9_12,

on the job, without the need for much additional support or education. These beliefs are prevalent because of our natural tendency to think in dichotomies, in this case 'content' vs 'pedagogy'. Those favouring one side of a dichotomy emphasize the importance of their activities (and sometimes of themselves) while downplaying the contributions of the other. However, in the real world of teachers, content and pedagogy are bound together; what one teaches and how one teaches it are part of the same package, not separate entities. As Dewey reminds us, it takes an "effort of thought" to see how apparently opposing factors in an educational situation can be brought together (Dewey, 1902, p. 4).

The need to unpack the meanings of 'content' and 'pedagogy' in order to bring them together is now widely recognized by most maths educators. As indicated in the first chapter, there is a growing number of studies, and a proliferation of distinctions and acronyms, that help to provide a nuanced understanding of what is meant by mathematical 'content'. Similarly, there is an increasing number of studies that focus on effective mathematics teaching by describing pedagogical approaches that engage learners and lead to desirable outcomes (Anthony & Walshaw, 2009; Martin, 2007). Some authors try to connect content and pedagogy. There are a number of interesting studies that relate content to pedagogy, for example by looking for a relationship between teachers' mathematical knowledge for teaching (MKT) and their decisions and actions (Charalambous, 2010) or by organizing maths curriculum and pedagogy around mathematical processes rather than products (Cuoco, Goldenberg & Mark, 2010). In *Adding it up*, the National Research Council explicitly characterized three kinds of knowledge that are crucial for teaching school mathematics: knowledge of mathematics, knowledge of students, and knowledge of instructional practices (NRC, 2001). They then go on to discuss, in detail with many helpful examples, how these three bodies of knowledge, taken together, can help us to develop proficiency in teaching mathematics. However, we argue that what is missing is another kind of knowledge: teachers not only need to know about mathematics and about their students, they also need to know about themselves, including their wishes, fantasies and habitual responses.

In this book, we have taken a different, more inclusive perspective. All these bodies of knowledge need to be connected. While it is certainly possible to study each factor separately for analytical purposes, there is a cost. In practice, *what* one teaches, *whom* one teaches, *how* one teaches, and the *self* of the teacher are part of a package that cannot be broken apart. Thus all of the chapters in this book centre on the teaching and learning of a single subject – mathematics – and each author clarifies what they do with this subject. In contrast, many self-studies are not bound to particular disciplines, electing to discuss teacher education practice in general. But, as Lighthall points out, "a narrative of the author's experiences with a given subject matter together with study of one or more pedagogical practices... [provides] rich insights" (Lighthall, 2004, p. 196). Insights also arise when we include knowledge of ourselves in the study, something each chapter also does. Hamachek, while arguing that knowledge of who we are is central to our teaching practice, says "Consciously, we teach what we know; unconsciously, we teach who we are" (1999, p. 209). In theory, we can distinguish these four factors from each other, but by separating the inseparable we run the risk of becoming less relevant to practising teachers who must deal with all four factors simultaneously.

There is another aspect of our perspective that deserves special mention. When discussing whom they teach, all of the chapters in this book pay particular attention to *specific* students and the data collected are almost always about *individuals*. Thus, when authors refer to the behaviour, understandings, or motivation of students, they are rarely talking about students in general. In contrast, in other books 'students' usually refers to collectives of students or to students in general. In *Adding it up*, for example, knowledge of students includes "general knowledge of how mathematical ideas develop. . . familiarity with the common difficulties. . . [and] knowledge about learning" (NRC, 2001, pp. 371–372). Such global generalizations are certainly useful for teachers (and are natural to find in a book of this nature), but they can only point the way. They are no substitute for the intimate and particular data that teachers need to collect.

Introducing Notions of the 'Third Space'

As we thought about the chapters in this volume, we came to see maths educators as occupying a unique third space, poised between the demands of their university courses and the exigencies of their students' future or current classrooms. The term 'third space' usually refers to out-of-school learning that occurs in social contexts. The first space is the place where formal learning happens. The second space – which can be a museum, or a library, or a home – is a site where informal learning takes place. In this context, the third space is the space that lies somewhere between these two major sites at which learning occurs. Oldenburg (2000) discusses the third space as the basis of community and social life and the space in which more creative activity can take place. Gutiérrez (2008) suggests it is the social context for learning that exists at the nexus between the formal and informal. It involves the re-imagining and recreation of those other spaces to provide an opportunity for creativity, insight and action.

This last sentence captures the essence of the metaphor and shows how the concept of a third space might be useful to us as we think about self-study in mathematics education. The third space, the intersection of different worlds or places with particular characteristics, becomes an interesting metaphor to examine. Although, like all metaphors, this one has limitations, if it can help us to re-imagine and recreate the first and second spaces, it might help us to bridge the apparent gap between inseparable factors in the situation. By providing an opportunity for creativity, insight and action, it will have served its purpose.

What then, in the context of mathematics education, might these spaces be? We see numerous possibilities. Most obviously, one space might be the 'mathematics space' and another, the 'teacher education space'; mathematics teacher education would be a third space in the intersection of these spaces. Or we could locate the third space between our selves as educators and the mathematics content that we teach, by viewing one space as being the so-called objective content of mathematics education and the other space as being our subjective experiences as mathematics educators. The third space in this example is the space in which our examination of our practice takes place; in this space we recognize that "how we teach is the message" (Russell, 1997, p. 32) that our teacher education students receive, and that the content and approaches that we use and share with our students are very much influenced by our own beliefs, experiences and attitudes.

Another third space lies between the two domains of mathematics and education. The two disciplines have different demands on their students. Fundamental to being a good mathematician is the ability to abstract, to use inductive and deductive reasoning and to generalize and extend results. Being a good teacher seems to demand related but distinct skills: the ability to communicate with your students, have a sense of what they believe, understand and can do, a caring disposition and an ability to enthuse and inspire. Mathematics education lies at the intersection of these two domains. As mathematics educators we strive to teach our students and student teachers how to develop mathematical habits of mind and also how to teach in ways that will enthuse future students with a love of mathematics, its elegance and its functionality. We aspire to provide our students with opportunities to grow in confidence in their mathematical endeavours, and to understand how to support their students to similarly grow. Thus the third space we inhabit here is located in the overlap between the development of ourselves as teachers and the content that we teach. When, as often happens, the first and second spaces have different features and goals, teachers face objective dilemmas and subjective tensions. As the chapters in this book illustrate, these dilemmas and tensions can be productively negotiated in the third space.

Locating the Book's Chapters in This Third Space

We can locate all the chapters in this book in the third space as defined above, as they all consider one or more aspects of that space. Paul Betts' chapter indicates the dilemmas and tensions that arise when he interrogates the value of MKT for different contexts, and he examines how his own conceptions grew through his interactions with other learners. Guðjónsdóttir and Kristinsdóttir examine the space between two worlds, the world of the special educator and the world of a mathematics educator. Cynthia Nicol describes the space between her own understandings about mathematics and those of her students. Sandy Schuck explores the space between her perceptions of her work in a primary teacher education program and the perceptions of a colleague who is not a mathematics educator. Robyn Brandenburg shows how attention to the space in which her students experience maths has enhanced her understanding and challenged her assumptions. Nell Cobb investigates the contrast between her understanding of fraction concepts and those of the teachers in her courses, and then describes ways that have helped her bring these two spaces together. Joanne Goodell is concerned about the habits of mind that will sustain new teachers in the face of resistance from their colleagues. Sue Gordon talks about the incongruities between her space and the spaces of her students that can occur when one uses metaphors to gain understanding of statistical concepts. Peter Pereira discusses the discrepancies between the space he inhabits and the spaces of middle school mathematics teachers. Ann Prescott asks how a secondary mathematics teacher educator can manage the tensions inherent in the process of becoming a reform based, effective mathematics teacher.

In all these cases, the intersection that is the self-study of mathematics education draws from both the first and second spaces and in so doing creates a third space with other characteristics. For us, the third space is the place of our activity as mathematics educators. It is also the place in which we can develop our understanding of the subjective experiences of teaching mathematics and the objective parts of the content and context in which we might teach. Finally, it is the place that we occupy as a field of endeavour as mathematics educators, where the first and second places might be our disciplines of mathematics and of teacher education. In all of these aspects of the third space, there are opportunities for re-imagining and creativity.

The content of mathematics discussed in this book varies from study of statistical concepts to understanding of fractions. Concepts such as Mathematical Knowledge for Teaching (MKT) (Ball, Thames & Phelps, 2008) and pedagogical content knowledge (Shulman, 1986) are relevant here. The content knowledge of teachers is critical, but, equally important, we must attend to students' struggles. This is where the self comes into the picture. As illustrated throughout this book, the authors are all engaged in developing, interrogating and critiquing the mathematical knowledge they believe is essential for teaching. Their passions, tensions, life histories and beliefs all come into play in their activities in constructing this content.

The Value of the Metaphor of the Third Space

What makes this idea of the third space interesting is the capacity that it has for re-imagining or revisualising our practice. Our self-studies do this by providing us with opportunities to individually or collaboratively interrogate our own practices and consider how they might look if we challenge our underlying beliefs and question our approaches. This encourages a reframing of our practice which takes the disparate contexts into account and develops a way of melding the features of those contexts into a way of guiding and enhancing our practice as mathematics teacher educators. Cobb provides us with an excellent example of how her self-study has illuminated her assumptions and created dissonances that promoted revisualizing of her practice as a mathematics teacher educator. Gordon, too, suggests that incongruities, which might have gone unnoticed had she not engaged in self-study, are unsettling her and providing opportunities to reframe her practice.

Another characteristic of the third space is its encouragement of creativity. Self-studies benefit from such creativity. This creativity can occur in the ways we collaboratively interrogate our practices, such as through the development of

Reflective Roundtable Inquiry as proposed by Brandenburg or through a collaborative self-study group as discussed by Nicol in her chapter, or through the participation of a critical friend from another discipline as in Schuck's chapter. The creativity can occur through the practices themselves, as indicated in Gordon's teaching using metaphor, or Guðjónsdóttir's and Kristinsdóttir's development of mathematics activities accessible to all. In these examples we see how creative practices and creative examination of those practices provide new ideas and challenges to our thinking about teaching of mathematics.

The third space is a place where we can consider how our self-studies contribute to our teaching and to our understanding of content and how best to teach it. Prescott's chapter illustrates this well as she examines the tensions she and her students experience and how this arises from their different perspectives, life histories and beliefs about mathematics. Her examination of these tensions indicates ways that she can manage or resolve them to better prepare her students for secondary school teaching. Betts describes how he was able to learn alongside the teachers with whom he worked. His mathematical knowledge for teaching was able to grow and change by working within this third space. Pereira discusses ways to help teachers change their understandings about mathematics thereby reinventing themselves as learners of mathematics. Goodell examines the impact of her teaching on the students' philosophies of mathematics teaching and gains valuable insights into her own practice by doing this.

Some Valuable Questions

The term 'tension' comes up in many places and is an important aspect of the book as a whole. It is a good word to use because it underscores the uncertain, problematic nature of teaching. Still, it is helpful to notice that it is used in several different ways. In Prescott's case it indicates a difficulty she faces in her practice that can be overcome. In Bett's case, the term is more problematic. In Pereira's case, tensions have an internal dimension as well as an external reality. Thus they never get resolved; our task is to negotiate our way through them rather than eliminate them. Like many things in life, they are always with us.

The recognition that the tensions we experience have an internal, subjective side as well as an external, objective component leads to some valuable questions that have application for other maths educators engaged in self-study. For example, Brandenburg asks, "Do I over-talk when I am trying to protect my identity, when I feel vulnerable?" This question is one all teachers need to keep in mind. Often, as teachers, we find it nerve-wracking to wait sufficiently long to get a response from our students. What happens if no one answers? What would that tell me, and my students, about myself as a teacher? Do we over-talk to cover the silence and the discomfort it provides?

Gordon poses another thought-provoking question by interrogating her own motives. When students tell her their short-term goals, she asks "Was my rejection or even denial of some students' instrumental or short-term goals based on my own need to transform students' outlooks rather than on pedagogical principles?" This question is particularly pertinent in maths education, where there is often a large gap between students' motives for studying a subject and our motives for teaching them. Students are often reluctant to be in mathematical or maths education subjects, completing them only because they are compulsory or because they realize that their career pathway will be blocked if they do not successfully complete them. Maths teachers not only want their students to have a successful passage through their courses, they want to infuse them with a love of maths and an appreciation of its beauty, elegance and functionality. We are often disappointed by the way students choose shortcuts over deep engagement. Gordon's question is a reminder to us that we need to continually investigate our motives in this regard. As she says, sometimes she had "to let go of the view that I am the sole owner of statistical knowledge and the overriding expert".

Gordon's awareness of her motives leads to discussion of another concept that is frequently questioned in this book, that of 'control'. Schuck notes the debates she and her critical friend have had regarding this issue. She feels that the need to question this aspect of our roles as teachers is particularly important in maths education. A number of other authors raise the issue of control and describe their internal conflicts regarding their desire to be 'the one in control'. Pereira tries to support his students, who as teachers are struggling with this notion. He asks them "Why do we have this need to feel in control? Why do we have so much trouble in giving it up?" Betts learns that he cannot always be the one in control. We all worry about losing control, although we vary in how we respond to this concern. What do we do when events threaten to get out of control? Do we make sure everyone knows who is in charge, or do we retreat behind the barrier of our mathematical competence? Do we put ourselves at the centre of the lessons, or do we allow our students to take that position? If we allow the latter, what implications are there for our control of the class and the learning?

These questions link to ones Betts poses about his internal motivations. He asks, "When do I tell, interact, respond? When do I critique, support, coach, guide, instruct? When am I learner, mentor? Will MKT be a barrier or a tool? When do I admit to inadequacy?" These questions will have a ring of familiarity about them to readers who study their own practice. The question of how much telling we should be doing and how much co-learning, of whether our role is as teacher or facilitator, is asked frequently in the literature and in practice. There is no one answer, but thinking about the options is likely to enhance our teaching. Pereira asks a similar set of questions, "Should I intervene or should I allow teachers to figure things out for themselves? Should I try to reduce frustration or is a certain amount of frustration a necessary part of learning?" He notes that withholding answers from students is often frustrating to them and realizes through his study that there is no one clear route through the obstacles and barriers of teaching and learning mathematics. The commonality in these questions indicates that those engaged in self-study of maths education understand this study to be filled with objective dilemmas and subjective tensions.

The Way Forward

In the above section we indicate the difficulty of finding answers in our self-study of teacher education practices in maths education. By implication we are indicating that the destination is not as important as the journey. The process of challenging our assumptions and beliefs and of striving to reframe our practices as we reinvent ourselves as teachers is inherently valuable. Doubt is a powerful driver of improvement. Schuck suggests in her chapter that she sees "complacency as an enemy of quality teaching". If this is true, then posing and trying to answer questions such as those above should be an essential part of our quest for quality teaching.

Still, lest you, the reader, feel somewhat cheated because we have managed to pose many questions but few solutions, there are answers to be found in the pages of this book. For example, Guðjónsdóttir and Kristinsdóttir suggest that their self-study has influenced the development of their *Mathematics for All* course, and that the collaboration between a special needs educator and a primary maths teacher has helped them to teach teachers who cater for a diverse range of students. Cobb has found that she has learnt much from investigating her students' understandings of fractional representation and that 'accurate empathy' and 'intellectual flexibility' strengthen her delivery of instruction. Prescott suggests a number of ways to resolve the tensions created by the differing perspectives of her students and herself. Betts has learned to accept that "that we cannot control teacher learning; that professional learning is often unexpected and accidental, and so all we can learn is what we happen to attend to". Nicol uses a collaborative group to share experiences and provide advice, while Goodell is currently embarked on the development of a new course that arose from her awareness, through self-study, of the need for improvement. Each of the authors has answered some questions while raising others.

Teachers, including maths educators, constantly have to reinvent themselves in order to stay fresh, competent and committed. To stand still would be to go backwards as competing demands, not to speak of the forces of nature, gradually undermine our efforts. We have proposed self-study as an activity that can rejuvenate our practice, and each of the chapters in this book illustrate ways that this could be done. Self-study, especially when done in collaboration with others, has helped us to reframe our practices and challenge our assumptions. We hope that this book encourages other mathematics educators to do the same. We face some big obstacles that we need to recognize, but that does not mean we should leave the field or take flight in various directions. We urge you, the readers, to explore your own third spaces as you consider what you teach, whom you teach, how you teach, and who you are.

References

Anthony, G., & Walshaw, M. (2009). *Effective pedagogy in mathematics*. Geneva: International Bureau of Education.

Ball, D. L., Thames, M. H., & Phelps, G. (2008). Content knowledge for teaching: What makes it special? *Journal of Teacher Education*, *59*(5), 389–407.

Charalambous, C. Y. (2010). Mathematical knowledge for teaching and task unfolding: An exploratory study. *The Elementary School Journal, 110*(3), 247–278.

Cuoco, A., Goldenberg, E. P., & Mark, J. (2010). Contemporary curriculum issues: Organizing a curriculum around mathematical habits of mind. *The Mathematics Teacher, 103*(9), 682–692.

Dewey, J. (1902). *The child and the curriculum.* Chicago: The University of Chicago Press.

Gutirrez, K. (2008). Developing a sociocritical literacy in the third space. *Reading Research Quarterly, 43*(2), 148–164.

Hamachek, D. T. (1999). Effective teachers: What they do, how they do it and the importance of self-knowledge. In R. P. Lipka & T. M. Brinthaupt (Eds.), *The role of self in teacher development* (pp. 189–224). Albany, NY: State University of New York Press.

Lighthall, F. (2004). Fundamental features and approaches of the s-step enterprise. In J. J. Loughran, M. L. Hamilton, V. K. LaBoskey, & T. Russell (Eds.), *International handbook of self-study of teacher education practices* (pp. 7–40). Kluwer: Dordrecht.

Ma, L. (1999). *Knowing and teaching elementary school mathematics: Teachers' understanding of fundamental mathematics in China and the United States.* New York: Lawrence Erlbaum Associates.

Martin, T. S. (Ed.). (2007). *Mathematics teaching today: Improving practice, improving student learning* (2nd ed.). Reston, VA: National Council of Teachers of Mathematics.

National Research Council. (2001). *Adding it up: Helping children learn mathematics.* Washington, DC: National Academy Press.

Oldenburg, R. (2000). *Celebrating the third place: Inspiring stories about the "great good places" at the heart of our communities.* New York: Marlowe & Company.

Russell, T. (1997). Teaching teachers: How I teach IS the message. In J. Loughran & T. Russell (Eds.), *Teaching about teaching: Purpose, passion and pedagogy in teacher education* (pp. 32–47). London: Falmer Press.

Shulman, L. S. (1986). Those who understand: Knowledge growth in teaching. *Educational Researcher, 15*(2), 4–14.

Author Index

S. Schuck, P. Pereira (eds.), *What Counts in Teaching Mathematics*, Self Study of Teaching and Teacher Education Practices 11, DOI 10.1007/978-94-007-0461-9,

Subject Index

MIX
Papier aus verantwortungsvollen Quellen
Paper from responsible sources
FSC® C105338

If you have any concerns about our products,
you can contact us on
ProductSafety@springernature.com

In case Publisher is established outside the EU,
the EU authorized representative is:
Springer Nature Customer Service Center GmbH
Europaplatz 3, 69115 Heidelberg, Germany

Printed by Libri Plureos GmbH
in Hamburg, Germany